3

AGRICULTURE ISSUE

U.S. NUTRITION AND AGRICULTURAL RESEARCH

OVERVIEW, FUNDING ISSUES AND TRENDS

AGRICULTURE ISSUES AND POLICIES

Additional books in this series can be found on Nova's website under the Series tab.

Additional e-books in this series can be found on Nova's website under the e-book tab.

AGRICULTURE ISSUES AND POLICIES

U.S. NUTRITION AND AGRICULTURAL RESEARCH

OVERVIEW, FUNDING ISSUES AND TRENDS

ARLENE RAMOS
EDITOR

New York

Library of Congress Cataloging-in-Publication Data

ISBN: 978-1-63484-170-2

Published by Nova Science Publishers, Inc. † New York

CONTENTS

Preface **vii**

Chapter 1 Improving Health Through Nutrition Research: An Overview of the U.S. Nutrition Research System **1**
Andrew A. Toole and Fred Kuchler

Chapter 2 Agricultural Research: Background and Issues **43**
Jim Monke

Chapter 3 Public Agricultural Research Spending and Future U.S. Agricultural Productivity Growth: Scenarios for 2010-2050 **61**
Paul Heisey, Sun Lin Wang and Keith Fuglie

Index **71**

PREFACE

This book explores the structure and function of the U.S. nutrition research system, with an emphasis on trends in Federal support. It describes how nutrition research is used, especially for nutrition education and communication, but also for regulation and food assistance. The book uses the Human Nutrition Research Information Management database maintained by the National Institutes of Health to analyze long-term trends. Furthermore, this book provides a background, and examines issues of agricultural research; and shows scenarios of public agricultural research spending and future U.S. agricultural productivity growth.

In: U.S. Nutrition and Agricultural Research ISBN: 978-1-63484-170-2
Editor: Arlene Ramos

Chapter 1

IMPROVING HEALTH THROUGH NUTRITION RESEARCH: AN OVERVIEW OF THE U.S. NUTRITION RESEARCH SYSTEM*

Andrew A. Toole and Fred Kuchler

ABSTRACT

This report explores the structure and function of the U.S. nutrition research system, with an emphasis on trends in Federal support. It describes how nutrition research is used, especially for nutrition education and communication, but also for regulation and food assistance. The report uses the Human Nutrition Research Information Management database maintained by the National Institutes of Health to analyze longterm trends. Federal investments in nutrition research grew from 1985 to 2009 in real terms, but the portfolio of research changed. Over time, the share of nutrition research support by the U.S. Department of Health and Human Services increased while support by the U.S. Department of Agriculture decreased. This shift changed how research topics were selected and funded within the Federal portfolio. More research is funded through competitive grants than through intramural or

---

* This is an edited, reformatted and augmented version of Economic Research Report Number 182, issued by the U.S. Department of Agriculture, Economic Research Service, January 2015.

formula funding, and a broader set of academic institutions now participates in nutrition research.

Keywords: Federal research, research and development, public investment, nutrition information system, Human Nutrition Research Information Management (HNRIM), formula funding, competitive funding

What Is the Issue?

A 2013 survey of consumers found that 45 percent were "very interested" and another 42 percent were "somewhat interested" in learning more about foods that have health benefits. Private companies are responding by redirecting their research and development efforts toward creating nutritionally enriched conventional foods and new food products that go beyond basic nutrition. Governmental organizations use programs and policies to address a variety of public health challenges such as malnutrition and obesity. The United States has a diverse and multidisciplinary nutrition research system with numerous sponsoring organizations and thousands of active researchers. The body of knowledge this system produces serves as the foundation for progress toward better health. However, maintaining this foundation is not guaranteed; it depends on the resources provided by sponsoring organizations and on how research topics are selected and funded. This report explores the structure and function of the U.S. nutrition research system, with a particular emphasis on changes in Federal support.

What Did the Study Find?

Public data on the levels and trends in research investments are limited, particularly for private research organizations. However, data are available on federally supported nutrition research from the Human Nutrition Research Information Management (HNRIM) database maintained by the National Institutes of Health. Analysis of the HNRIM database for the 25 years from 1985 through 2009 (the latest year of available USDA data) revealed that Federal investments in nutrition research more than doubled in real terms, but the portfolio of research changed. The share of Federal support by the U.S. Department of Health and Human Services (DHHS) increased while that by the U.S. Department of Agriculture (USDA) decreased. This shift changed how research topics were selected and funded within the Federal portfolio. As a result, more research is funded through competitive grants than through intramural or formula funding, and a broader set of academic institutions now participates in nutrition research.

To analyze the nutrition research areas receiving Federal support, we construct a portfolio that shows the shares (percent of all nutrition research projects) across 6 major research areas spanning 37 topics. The analysis finds:

- The *Disease, Injury, and Conditions* research area grew most in its share of Federal support, climbing from 40 percent in 1985 to 49 percent by 2009. This area covers a wide range of diseases (e.g., cardiovascular, diabetes, and cancer) and conditions (e.g., obesity, anorexia, and high cholesterol).
- Within the Disease, Injury, and Conditions research area, the *Obesity/Anorexia/Appetite Control* topic grew fastest, rising from 3.6 percent of the Federal portfolio in 1985 to 13.1 percent in 2009. This topic area grew fastest within the portfolios of both DHHS and USDA.
- The *Metabolism and Metabolic Mechanisms* research area experienced the largest decline in Federal portfolio share, from 28 percent in 1985 to 20 percent in 2009. This area investigates how the human body gets or makes energy from food and its constituents such as carbohydrates and proteins.
- Seven out of the nine topic areas that make up the *Metabolism and Metabolic Mechanisms* research area experienced declining Federal shares between 1985 and 2009. The topics leading this decline were *Vitamins, Minerals, and Proteins*.
- The USDA provides most of the Federal support for nutrition research in the *Food Sciences* area, which includes food processing, preservation, and other food-related technologies. From 1985 to 2009, USDA supported 80 percent of the active projects (on average).
- The Federal portfolio share allocated to *Food Sciences* decreased from 10 to 4 percent in the period analyzed. While the Federal shares for each of the four topic areas within the *Food Sciences* fell, the topics leading the decline were *Food Composition* and *Effects of Technology on Foods and Diets*.

The portfolio of federally supported organizations that perform nutrition research also shifted from 1985 to 2009:

- The share of Federal nutrition research projects performed by *Government researchers* fell from 12 percent in 1985 to 6 percent in 2009.
- The share of research projects performed by *land-grant universities and colleges* fell from 34 percent in 1985 to 22 percent in 2009.
- The share of research performed by *non-land grant universities and colleges* grew from 30 percent in 1985 to 41 percent of total nutrition projects in 2009.
- The *Other* category of institutions—*medical schools, hospitals, and research institutes*—also saw its share of research support grow, from 22 percent to 29 percent of all federally supported nutrition projects.
- *Private companies* performed just 1-2 percent of federally supported nutrition research projects.

Our review of the academic and policy literatures found no published studies that analyzed the impacts of these trends on the volume of research performed or the productivity of the U.S. nutrition research system. Future studies could analyze available indicators of research outputs, such as published articles or patents, and could relate those to inputs, such as project effort or financial investments.

How Was the Study Conducted?

This report synthesizes the existing literature analyzing the U.S. nutrition research system and analyzes data on Federal support from the Human Nutrition Research Information Management (HNRIM) database. This database is maintained by the National Institutes of Health (NIH) under the auspices of the Interagency Committee on Human Nutrition Research. Each participating Federal agency reports to HNRIM its own data on active nutrition projects each year. To document the nature and trends in Federal nutrition research, this report relies on the number and distribution of active project counts contained in HNRIM for fiscal years 1985 through 2009. Based on an analysis of available data on project award amounts, we concluded that project counts accurately characterize trends in Federal support for nutrition research.

INTRODUCTION

U.S. consumers, private companies, and governmental organizations are all keenly interested in the relationships between nutrition, dietary choices, and health. In 2013, a survey by the International Food Information Council found that 45 percent of consumers were "very interested" and another 42 percent were "somewhat interested" in learning more about foods that have health benefits (IFIC, 2013). Private companies are responding to consumers by redirecting their research and development efforts toward creating nutritionally enriched conventional foods and new food products that go beyond basic nutrition. Governmental organizations use programs and policies to address a variety of public health challenges such as malnutrition and obesity.

The links between nutrition, dietary choices, and health are established through research. *Basic research* reveals the genetic, cellular, and chemical relationships that determine metabolic responses to diet and food components. For instance, *nutrigenetics/nutrigenomics research* is expected to produce new nutritional and disease biomarkers based on how nutrients interact with genes, proteins, and metabolites for specific individuals (Ohlhorst et al., 2013). *Clinical research* translates basic research into evidence-based policy and practice. These studies are critical for demonstrating nutrient dosages, absorption, efficacy, and safety for health interventions. *Observational research* focuses on a variety of topics including nutritional epidemiology, nutrition education, and program evaluations. Insights into food consumption habits and behaviors from these studies help to improve food choices, policies, and regulations.

As a whole, this body of nutrition research underpins progress toward better health. However, maintaining and strengthening this knowledge foundation depends on resources provided by sponsoring organizations and on how research topics are selected and funded. For instance, changes in Federal budget appropriations across agencies affect not only the level of Federal support but also how Federal support is administered. When Federal agencies have different missions, use different funding mechanisms, and support different performers within the nutrition research system, these funding changes may alter the nature of nutrition research by favoring some topics and performers over others or affecting research productivity. This report provides a first look at the structure and function of the U.S. nutrition research system, with an emphasis on changes in Federal support.

Paths from Nutrition Research to Better Health

Nutrition research produces information that influences human health through a variety of paths within a complex communication system, composed of information generation, translation, and impact (fig. 1). Information generation, carried out by the nutrition research system, is the stage where researchers discover new knowledge, methods, and technologies. For example, in 2007, researchers identified the so-called "fat mass and obesity-associated" (FTO) gene variant, which is present in about one out of every six adults in the population. People with the FTO gene variant have a 20- to 30-percent higher risk of obesity compared to those without the variant (Harvard School of Public Health, 2014). Other researchers are using this information to understand how the FTO gene variant triggers weight gain. In a 2013 study, Rachel Batterham and her research team identified another link. She described it as follows:

> We've known for a while that variations in the FTO gene are strongly linked with obesity, but until now we didn't know why. What this study shows us is that individuals with two copies of the obesity-risk FTO variant are biologically programmed to eat more. Not only do these people have higher ghrelin levels and therefore feel hungrier, their brains respond differently to ghrelin and to pictures of food—it's a double hit. (University College London, 2013)

Discoveries like the FTO gene variant are communicated through a variety of channels, such as publications, conferences, interpersonal networks, and consulting, to the next component—information translation. Information translation is essential because consumers rarely draw information directly from the scientific journals or other communication channels used by nutrition researchers. Instead, consumers rely on various public and private organizations to evaluate, interpret, and repackage nutrition research findings to facilitate understanding and guide behaviors. These translational activities include public-sector communication of best practices, public-sector regulations and policies, and private-sector products and services (fig. 1). The connection between nutrition research and each of these translational activities is discussed below in separate subsections.

The findings from nutrition research need to be accessible and understandable before consumers can use this information to improve their choices about food and diet.

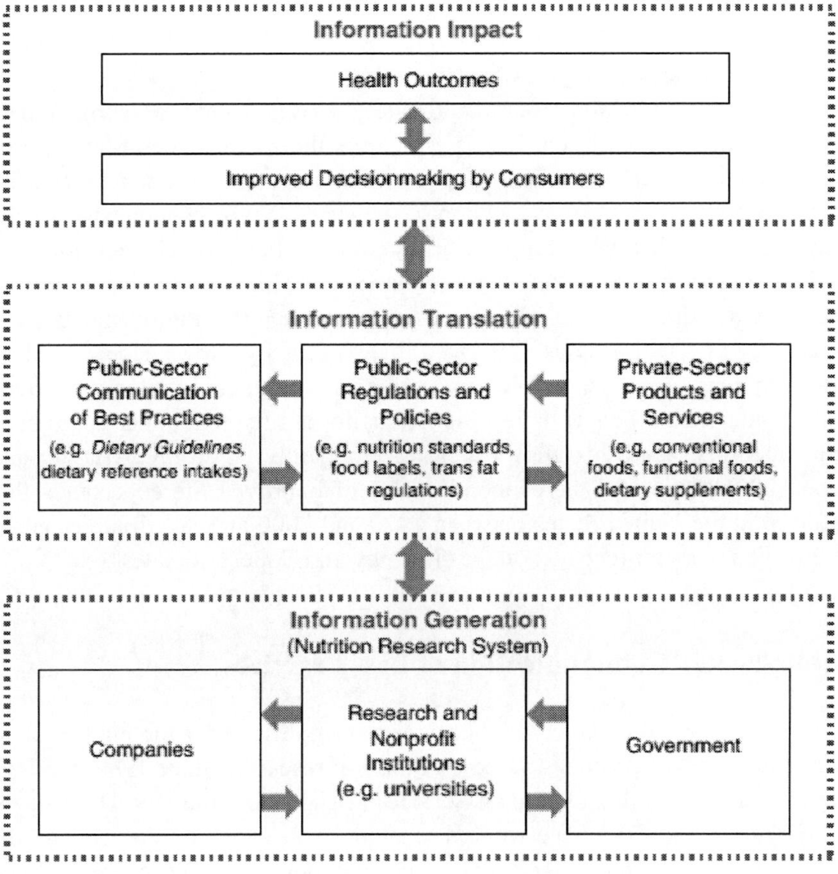

Source: USDA, Economic Research Service, author's calculations.

Figure 1. Schematic of the nutrition information communication system.

Consumer choices, in turn, influence short-term and long-term health. For instance, dietary choices affect the extent and severity of diseases such as cardiovascular disease and conditions such as anemia, anorexia, and obesity. The trends and costs associated with nutrition-related diseases and conditions are significant concerns within the public health community. Obesity rates among adults increased by 37 percent between 1998 and 2006, with an obese adult spending $1,429 more on health care than a normal-weight person in 2006 (Finkelstein et al., 2009). And these obesity-driven increases in health care costs are likely to continue. After a review of recent epidemiological studies, the Centers for Disease Control and Prevention (CDC) concluded that "children and adolescents who are obese are likely to be obese as adults" and "childhood obesity has more than doubled in children and quadrupled in adolescents in the past 30 years" (CDC, 2014).

The value of nutrition research ultimately derives from its potential impact on human health. While the health effects of the research described in this report have not been quantified, it is clear that even a small impact on health would produce very large economic benefits. Two economists at the University of Chicago, Kevin Murphy and Robert Topel, developed a framework for understanding the economic value created by better health. Their framework considered two aspects of health improvements—life expectancy and health status. Improved life expectancy allows people to enjoy more products, services, and leisure activities over their lifetime. Improved health status means people have a higher quality of life (e.g., better eyesight or better mobility), and this allows them to more fully enjoy life. Murphy and Topel (2006) estimated the economic value of improved life expectancy that occurred in the United States between 1970 and 2000 at $95 trillion, or about $3.2 trillion per year (see box on the "Murphy and Topel Estimates").

Public-Sector Communication of Best Practices

The *Dietary Guidelines for Americans* exemplifies how the public sector communicates best practices based on nutrition research. Since 1980, the U.S. Department of Health and Human Services (DHHS) and the U.S. Department of Agriculture (USDA) have worked together to develop and disseminate the *Dietary Guidelines for Americans*.[1] The *Guidelines*, released every 5 years, provide recommendations for a nutritionally balanced diet developed from a review of relevant scientific evidence.

Murphy and Topel Estimates: Perspective and Approach to Health Valuation

The $95-trillion economic gain calculated by Murphy and Topel is enormous. To put this in perspective, Murphy and Topel compared their figure with the gross domestic product (GDP) of the United States. The U.S. GDP measures the market value of all final goods and services produced by labor and property located in the United States. Economists and policymakers use the GDP to assess the levels and trends in U.S. output and economic growth.

Similarly, Murphy and Topel view annual health improvements as health "output" that adds to the wellbeing of Americans. Making this comparison, the $95 trillion Murphy and Topel calculated from increased life expectancy over 1970 to 2000 is equivalent to 44 percent of the cumulative GDP for the United States over the same period. In other words, the value afforded Americans through improved life expectancy was equal to 44 percent of the total value of all final goods and services produced in the United States over those years.

Why are the estimated values for health improvements so large? As Murphy and Topel point out, the economic gains are large because the number of people who benefit from the health improvements is large. Murphy and Topel started with the observation that during the 20th century, life expectancy at birth for a typical American increased by 30 years. The value of greater longevity and better health, however, is not like the value of cars, toasters, or other things that might be purchased in retail stores at a known price. Instead, the value of greater longevity and better health must be estimated indirectly. Murphy and Topel based their valuation on how much a person would pay (called "willingness to pay") for a risk reduction, where the risks considered influence the length and quality of the person's life.

Of course, measuring the value of risk reduction is not easy compared to measuring the value of other product attributes that are priced in the market. For example, Coca-Cola soft drinks are usually more expensive than generic colas of the same size. This price premium for Coca-Cola indicates the value consumers place on the Coca-Cola brand. Product labels make it easy for consumers to compare prices and many product attributes, but few products indicate the degree of riskiness. For instance, no one sells chicken parts with labels indicating the likelihood of contracting a *Salmonella* infection. Peoples' willingness to pay for risk reduction, however, can be detected in some market transactions, and economists use those transactions to estimate the value of risk reductions. The standard approach is to look at job market transactions, although product market transactions (e.g., purchases of bicycle helmets) and housing market transactions (e.g., home purchases in locations at risk for a hazardous waste site) are also used in some studies. In the job market, people are generally paid more to perform riskier jobs, and economists have estimated how much wages rise when workers accept a small risk of on-the-job fatal injury using information from jobs in mining, construction, manufacturing, and many other industries (Viscusi and Aldy, 2003).

From this, economists estimate the risk-dollar tradeoffs individuals are willing to accept.

Murphy and Topel used risk-dollar tradeoffs from job market studies to estimate a yearly value of a “statistical life” at every age. They benchmarked their analysis to studies estimating that workers demand a $630 annual wage premium to accept a 1-in-10,000 risk of fatal injury. That is, across 10,000 workers accepting the wage premium and the risk, we expect to see one fatal injury for an increased wage bill of $6.3 million. In their analysis, $6.3 million is the average value of a statistical life for working-age adults, a figure obtained from the U.S. Environmental Protection Agency. Based on this, Murphy and Topel created an annual version called the “value of a life-year” for an individual at every age, beginning at birth and extending to death as determined by the mortality tables for the United States.

Using the observed increase in life expectancy in the United States, Murphy and Topel calculated the increase in the value of a life-year for an individual. Then, for individuals at each age, they summed the increased life-year values over the remaining years in the person’s lifespan. This produces a total increase (or gain) in the value of an individual’s statistical life at each age. They still have two more steps in their calculation. First, to calculate the economic value across all people in an age group, they must multiply the increased value of life at each age by the population in that age group. They do this for the current population and for future populations as future generations are also expected to benefit from increased life expectancy. This step dramatically increases their estimated gains because there are millions of people in the population at each age. Now, with the total benefits for each age group, the final economic value is reached by summing across all the age groups in the current population as well as those expected to benefit in future populations.

The science-based process of creating the *Guidelines* begins with the selection of an independent Dietary Guidelines Advisory Committee (DGAC) composed of nationally recognized experts in the fields of nutrition and health. These individuals are responsible for making recommendations to the Secretaries of DHHS and USDA after a thorough review of current scientific and medical knowledge. The DGAC uses the Nutrition Evidence Library (NEL) to support its systematic review. The NEL reviews, evaluates, synthesizes and houses the information produced by the nutrition research system and posts full “evidence portfolios” online (see http://www.nel.gov/).

The *Guidelines* are the foundation for educational materials like *MyPlate* distributed by the USDA and other Federal agencies. By law,[2] all Federal dietary guidance for the public—including print and Web-based educational materials, messages, tools, and programs to communicate healthy eating and physical activity information—must be consistent with the *Guidelines*.

But the influence of the *Guidelines* does not end with advice. The *Guidelines* influence the level and type of food assistance offered through the 15 food and nutrition assistance programs administered by USDA's Food and Nutrition Service. USDA's Center for Nutrition Policy and Promotion relies on the *Guidelines* as the nutritional basis for its *USDA Food Plans* (Thrifty, Low-Cost, Moderate-Cost, and Liberal). The Thrifty Food Plan is an estimate of how much it costs to buy food and to prepare nutritious, low-cost meals for a household. That estimate is periodically revised for consistency with the *Guidelines* and is the basis for household allotments under the Supplemental Nutrition Assistance Program (SNAP–formerly the Food Stamp Program). The *Guidelines* are an input into determining food benefits under the Special Supplemental Nutrition Program for Women, Infants, and Children (WIC).[3] And the *Guidelines* influence meal standards for the National School Lunch Program.

Public-Sector Regulations and Policies

Nutrition research is also an integral source of information for regulations and policies such as the Federal regulations on retail food labels. Under the Nutrition Labeling and Education Act (NLEA) of 1990, the U.S. Food and Drug Administration (FDA) establishes how food manufacturers will account for nutrients in foods—the Nutrition Facts panel on processed foods—and modifies those standards periodically (see box, "Trans Fats Labeling"). Another example is USDA's nutrition standards for all foods sold in schools, including snacks and vending machine foods that compete with federally supported meals programs. The Healthy, Hunger-Free Kids Act of 2010 requires USDA to establish nutrition standards for all foods sold in schools—beyond the federally supported meals programs. The *Smart Snacks in School* nutrition standards drew on recommendations from the Institute of Medicine and voluntary standards already implemented by schools around the country (FNS, 2013).

Nutrition Research Supports Federal Food Labeling Regulations— Trans Fats Labeling

While the use of trans fats in food manufacturing became widespread in the 1940s (DHHS, FDA, 2013), their use is now declining (Rahkovsky et al., 2012). Trans fats were commonly used in food processing because hydrogenation allowed food processors to raise the melting point of relatively inexpensive products containing polyunsaturated fatty acids, making the products solid at room temperature.[1] Foods that are higher in saturated fatty acids are more resistant to spoilage. The twin attractions of these synthetic trans fatty acids, found in partially hydrogenated oils, are reduced costs of food production and longer shelf life of products. Partially hydrogenated oils have frequently been used in margarines, snack foods, and prepared desserts, replacing saturated fatty acids.

Trans fats made it easier for food manufacturers to develop tasty snacks, and their snacks cost less to make than if they had used other saturated fats. Consumers likely shared in those benefits as the cost reductions kept retail prices of snack foods down. However, nutrition research identified a downside to the widespread use of trans fats in manufactured foods, as summarized by the Institute of Medicine (part of the National Academy of Sciences):

> Trans fatty acids are not essential and provide no known benefit to human health. Therefore, no AI or RDA is set. As with saturated fatty acids, there is a positive linear trend between trans fatty acid intake and LDL cholesterol concentration, and therefore increased risk of CHD. A UL is not set for trans fatty acids because any incremental increase in trans fatty acid intake increases CHD risk.[2]

The Federal Government has pursued two activities intended to reduce Americans' dietary intake of trans fats:

1. Informing consumers that they can reduce health risks by choosing fats other than trans fats, and
2. Requiring food manufacturers to label the trans fats content of foods.

Informing consumers

The recommendation to minimize intake of trans fats began with the *Dietary Guidelines for Americans*, 2005.[3]

> Consume less than 10 percent of calories from saturated fatty acids and less than 300 mg/day of choles terol, and keep trans fatty acid consumption as low as possible.

At that time, the Food Guide Pyramid emphasized the benefits of reducing foods high in trans fatty acids. The 2010 *Dietary Guidelines* offers additional advice on how to minimize trans fat consumption:

> Keep trans fatty acid consumption as low as possible, especially by limiting foods that contain synthetic sources of trans fats, such as partially hydrogenated oils, and by limiting other solid fats.

Regulations for food labels

In November 1999, the U.S. Food and Drug Administration (FDA) proposed a rule that would require the amount of trans fatty acids present in foods to be included in the product's Nutrition Facts panel. In 2003, the rule became final with requirements for labeling beginning in 2006.

In its November 1999 proposed rule, FDA described the nutrition research studies that it relied on. It had to establish that dietary intake of trans fats could be linked to adverse health outcomes—increases in LDL-C would imply a consequent increase in rates of coronary heart disease (CHD). FDA also had to gauge the magnitude of the problem, showing how much trans fats Americans were consuming. And it had to quantify how different types of Government intervention would improve health outcomes. FDA relied on studies published between 1988 and 1995 (p. 41442 in final rule). The most problematic studies were those examining the linkage between dietary intake of trans fats and CHD. These included controlled intervention studies, which tested for a causal relationship between trans fat consumption and CHD, and epidemiological studies, which established associations between trans fat consumption and CHD. FDA interpreted the results from published studies as "evolving." Some studies pointed to trans fats raising LDL-C levels (the primary risk factor for CHD) as much as cholesterol-raising saturated fatty acids, while others found no adverse effect. Between the November 1999 proposal and the July 2003 final rule, additional studies reinforced FDA's conclusion that trans fat consumption raised LDL-C levels and increased CHD risk.

FDA's estimates of the intake of trans fats were established by drawing on several sources, but largely came from USDA's Continuing Survey of Food Intakes by Individuals and from USDA-ERS figures on fats and oils production and food disappearance data (recognized as an overestimate).

FDA relied on its own survey of consumer behavior to gauge how much consumers might respond to finding trans fats on nutrition facts panels. While many studies link diet and health outcomes, behavioral studies of label use relevant to trans fats are still few.

In November 2013, FDA announced that additional scientific evidence led the agency to tentatively determine that partially hydrogenated oils—the primary source of trans fats in manufactured foods—are not generally recognized as safe for any use in food. If the tentative determination is supported, partially hydrogenated oils would be considered food additives and their use in food manufacturing subject to regulation.

Driving the tentative determination were new research findings. FDA's own studies pointed to trans fat consumption trending lower as foods were reformulated. However, new studies of health risks pointed to increased risk of CHD from any amount of trans fat. Other new studies found, with varying degrees of certainty, that trans fat consumption may worsen insulin resistance or raise diabetes risk. Clearly, as research results unfold, the regulatory process responds.

[1] Dietary Guidelines for Americans, 2010, Part D Section 3: Fatty Acids and Cholesterol, p. D3-7.

[2] AI, Adequate Intake, is defined as the recommended average daily intake level. RDA, Recommended Dietary Allowance, is defined as the average daily dietary nutrient intake level sufficient to meet nutrient requirements. UL, Tolerable Upper Intake Level, is defined as the highest average daily nutrient intake level that is likely to pose no risk of adverse health effects to almost all individuals in the general population. Coronary heart disease is abbreviated as CHD.

[3] Because trans fatty acids are unavoidable in ordinary, nonvegan diets, consuming 0 percent of them as energy would require significant changes in patterns of dietary intake. As with saturated fatty acids, such adjustments may introduce undesirable effects (e.g., elimination of commercially prepared foods, dairy products, and meats that contain trans fatty acids may result in inadequate intakes of protein and certain micronutrients) and unknown and unquantifiable

> health risks. Nevertheless, it is recommended that trans fatty acid consumption be as low as possible while consuming a nutritionally adequate diet. (Institute of Medicine, 2005, pp. 423-24)

Food policies, like requirements for iodine fortification in salt or the enrichment of flour products with B vitamins and iron, eliminated many nutritional deficiencies in the United States (Davis and Saltos, 1999). For instance, iodine deficiency impedes the proper functioning of the thyroid and is associated with a number of disorders including brain damage in children. In 1996, the FDA established regulations requiring that by 1998 all standardized enriched cereal grain products sold in the United States include 140 micrograms of folic acid per 100 grams, and allowed for the addition of folic acid to breakfast cereals, corn grits, infant formulas, medical foods, and foods for special dietary use. Before fortification, about 4,130 U.S. babies had neural tube defects each year and nearly 1,200 died. After folic acid fortification, the yearly number of pregnancies affected by neural tube defects dropped to about 3,000, and related deaths declined to 840 (CDC, 2004).

The notion that the Federal Government might solve or at least mitigate specific health problems by requiring the addition of micronutrients could only come from research demonstrating that such micronutrients are critical to avoiding a disease. However, a single research finding is unlikely to justify new regulations. Federal rulemaking and regulations are predicated on results from across the spectrum of nutrition research. Studies on nutrient metabolism, dietary intake, food consumption habits, and the effect of regulations on food choices are all critical links in demonstrating whether Federal intervention in dietary issues is worthwhile.

Private-Sector Innovation

As consumers demand healthier food and beverage products, the need for nutrition research to underscore these innovative products has grown. One approach to product innovation focuses on reformulating conventional food products to increase or decrease particular nutrients such as fat, sodium, and fiber. Companies can market these changes to consumers by using health- and nutrition-related claims, which are regulated by the FDA under the Nutrition Labeling and Education Act of 1990. In 2009, sales of products with sodium-, fat-, and calorie-related claims reached $73 billion (Martinez, 2013).

Another approach to product innovation, called "functional foods," is based on the idea that foods and drinks can provide health benefits beyond basic nutrition. No legal definition exists for functional foods in the United States, and the FDA regulates this type of food according to its classification as conventional food (e.g., garlic, nuts, and tomatoes), modified food (e.g., fortified, enriched, or enhanced), a food additive, a dietary supplement, a medical food, or a food for special dietary use (Academy of Nutrition and Dietetics, 2009). Functional foods are one of the fastest growing product categories in the food and beverage industries. According to Price Waterhouse Coopers (2009), U.S. revenues in the functional foods segment are growing at an average annual rate between 8.5 and 20 percent, whereas overall (food) industry revenue growth is at 1 to 4 percent.

The health connection that qualifies particular foods and beverages as functional foods is established through nutrition research. The Institute for Food Technologists outlined a process for bringing functional foods to market, whereby the nutrition research system (1) identifies the relationship between a food component and a health outcome, (2) demonstrates efficacy and intake levels, and (3) determines safety at efficacious levels (IFT, 2005). For instance, phytosterols are essential components of plant membranes that occur naturally in a variety of foods such as vegetables, fruits, and legumes. Nutrition research has shown that daily consumption of 2-3 grams will reduce serum LDL cholesterol and thus lower the risk of coronary heart disease (Demonty et al., 2009). Phytosterols are now added to many foods and beverages such as orange juice, cereals, yogurts, and granola bars.

THE U.S. NUTRITION RESEARCH SYSTEM: NATURE AND STRUCTURE

The nutrition research system is foundational in that it generates the new knowledge and technologies that fuel improvements in food, dietary choices, and health outcomes. As such, continuing scientific progress toward understanding the relationships between human nutrition and health depends on maintaining a productive nutrition research system. This section describes the nature of nutrition research, the institutions that support nutrition research, available data, and trends in Federal support.

The field of human nutrition research is diverse and multidisciplinary. In 1980, the Joint Subcommittee on Human Nutrition Research formed by the

Office of Science and Technology Policy defined nutrition research as those studies producing "new knowledge to improve the understanding of nutrition as it relates to human health and disease" (GAO, 1982). This definition includes disciplines such as food and plant sciences, biomedicine, behavioral science, education, communication, and economics. It cuts across the full research spectrum from basic science to health policy and from discovery to application.

The support for nutrition research activities comes from a variety of sources. In the private sector, profits motivate companies to innovate, undertake in-house research and development (R&D), and sponsor research in universities and government labs. Data on U.S. private R&D investments into nutrition research are not publicly available. However, private sector nutrition research is part of total R&D investments in the food and beverage industries, and we can gauge these latter investments. In 2007, Statistics Canada conducted a representative survey of Canadian companies active in the functional foods and natural health products markets (i.e., dietary supplements), which are the nutrition research-intensive market segments in the food and beverage industries. The survey indicated that 68 percent of the firms in the functional foods segment and 48 percent in the natural health products segment were engaged in R&D and spent a total of 148 million Canadian dollars in 2007 (Cinnamon, 2009). USDA's Economic Research Service estimated total R&D investments by the food and beverage industry at over $3.2 billion in 2006 (Fuglie et al., 2011). Even a small fraction of this total devoted to nutrition research would represent a substantial private-sector commitment.

Universities and other nonprofit organizations also support U.S. nutrition research. Here again, publicly available data are limited so it is difficult to draw precise inferences about relative magnitudes of financial support. The National Science Foundation's Higher Education R&D Survey is the main source for information on R&D conducted at U.S. universities and colleges. A breakout by academic discipline is not available, but the survey data are broken out according to the type of funding source. From 2001 to 2011, higher education institutions funded an average of 19 percent of total university- and college-performed R&D, and other nonprofit sources such as foundations funded an average of 7 percent (NSF, 2013). As human nutrition research is likely to be only a small fraction of total university R&D, the level of support provided by research and nonprofit organizations appears to be small relative to industry and government sources.

Federal and State governments appear to be the largest supporters of U.S. nutrition research, which makes sense from an economics perspective. Nutrition research produces new knowledge that economists associate with "public goods." With public goods, the incentives driving private companies to invest are small even when the benefits that consumers enjoy are large (Ribaudo et al., 2008). The problem is that private companies may not be able to earn enough profits from the products, services, and technologies that flow from their nutrition research investments to make such investments worth undertaking. For example, the genetic research that led to the identification of the "obesity gene" was not immediately or eminently marketable. Public support can offset this problem by funding research in areas where private companies are unwilling to invest (or invest very little) due to insufficient incentives.

Federal Support for Nutrition Research

Systematic data on Federal support for nutrition research and training are available from the Human Nutrition Research Information Management (HNRIM) system.[4] This database is maintained by the National Institutes of Health (NIH). Each participating Federal agency provides its own data on active nutrition projects each year to form the HNRIM database. Since 1985, when the HNRIM database became operational, participating agencies have included the U.S. Department of Health and Human Services (DHHS), the U.S. Department of Agriculture (USDA), and six other Federal agencies (the Department of Veteran Affairs, the Agency for International Development, the U.S. Department of Defense, the U.S. Department of Commerce, the National Science Foundation, and the National Aeronautics and Space Administration).

The HNRIM database includes records for active nutrition research and training projects in each fiscal year. A multiyear project has multiple records in the database, one for every year the project is active. Across time, new projects enter the database, and completed projects are no longer reported to the database. This change in the number of active projects is used to characterize the evolution of Federal support for nutrition research for fiscal years 1985 through 2009.[5] Over that 25-year period, HNRIM contains 100,405 project records.[6] Project records in HNRIM also include a nutrition percentage (from 1 to 100) that adjusts for the relevance of the project to the field of nutrition. When each annual project is scaled by its nutrition relevance, the total number of federally supported nutrition research projects in HNRIM for fiscal years 1985 through 2009 is 67,958. In this analysis, all project statistics are calculated and reported accounting for the nutrition percentages.

To document the nature of and trends in nutrition research and training, our analysis uses the number and distribution of active project counts contained in HNRIM.[7] Based on analysis of available data on project award amounts, we concluded that project counts better characterize Federal research trends (see the appendix for more details). Overall, the HNRIM data on financial awards are not comprehensive enough to perform a comparative analysis.

Other than DHHS, all Federal agencies that support nutrition research reported $0 as the financial support for a significant number of projects. For USDA projects, 15 percent show $0 as the funding amount. For the six remaining agencies, about 63 percent of the projects report $0 as the funding amount. It is possible for a project to be completed after its planned end date, and some projects operate for a time through no-cost extensions. Reporting $0 would be appropriate for no-cost extensions, but it is also possible that the financial data for these projects are missing from HNRIM.

Because no-cost extensions represent ongoing nutrition research and these projects cannot, with certainty, be distinguished from projects missing financial data, we present overall trends in Federal support using expenditures and project counts. When analyzing research area components, project counts are used.

National Trends in Federal Support for Nutrition Research

Over the last decade, the DHHS and USDA supported almost all Federal nutrition research projects. HNRIM data show that since 1998, the other 6 agencies reported 51 projects, or less than 0.2 percent of the total. In the mid-1990s, the other six agencies made up a more significant share. In 1993 and 1994, they accounted for 17 percent of active nutrition research projects, but by 2005 their share had dropped to zero. Although the HNRIM database does not show any active projects for the other six agencies after 2005, one cannot conclude that these agencies did not support any nutrition-related research. Participation in the HNRIM system is voluntary, and reporting may have stopped for other reasons. Nevertheless, as projects supported by agencies other than DHHS and USDA are a consistently small share of the total, for the remainder of this discussion we refer to the totals of DHHS and USDA as the Federal total, and our focus is on projects supported by DHHS and USDA.[8]

Federal financial investments into nutrition research more than doubled from 1985 to 2009, growing at an average annual rate of 3 percent (fig. 2). All of this growth is due to increased DHHS funding, especially between 1999 and 2003. In those 5 years, Congress implemented its plan to double the budget of the National Institutes of Health (NIH), which is the lead agency within DHHS supporting nutrition research. USDA funding fell at an average annual rate of 1.4 percent between 1985 and 2009.

When charted by active Federal nutrition projects, the overall Federal and agency-specific results are quite similar (fig. 3); however, the “kink” in the expenditures trend from the doubling of the NIH budget is no longer apparent. This suggests that increasing project size (i.e., dollars per project) absorbed a lot of the NIH budget increase. Nevertheless, the number of federally supported nutrition research projects more than doubled from 1985 to 2009, growing at an average annual rate of 2.9 percent. Nutrition projects numbered 2,178 in 1985 and 4,419 in 2009. This growth was the result of increasing support by DHHS and decreasing support by USDA. From 1999 to 2009, the number of DHHS-supported projects grew 7.4 percent annually while USDA-supported projects fell by 2.8 percent.

The aggregate trends in Federal support for nutrition research largely reflect budget appropriations to DHHS, particularly the NIH. As the NIH budget grew, the share allocated to nutrition research remained nearly constant and averaged about 4 percent per year from 1985 through 2009 (see HNRIM reports at http://hnrim.nih.gov/yearlyreports.aspx). For USDA, there may have been some reallocation of research effort. Since 1985, USDA public research expenditures remained nearly constant as the number of active nutrition research projects declined (http://www.ers.usda.gov/dataproducts/agricultural-research-funding-in-the-public-and-private-sectors.aspx). These trends tipped the balance of Federal support for nutrition research heavily toward DHHS. In the 25-year period covered by the data, the share of total Federal support provided by the USDA fell from 29 to 14 percent, on net, with a concurrent rise in the DHHS share to 86 percent by 2009.

As the balance of Federal support shifted toward the DHHS, the mechanisms used to administer Federal support also changed. Federal support mechanisms—the channels through which Federal agencies try to stimulate and advance human nutrition research—have a direct influence on the success of research activities and the accumulation of new knowledge. These mechanisms significantly influence the topics studied, the selection processes determining who gets support, the type of research activities (i.e., laboratory, clinical, social, etc.), and how research findings are communicated and used.

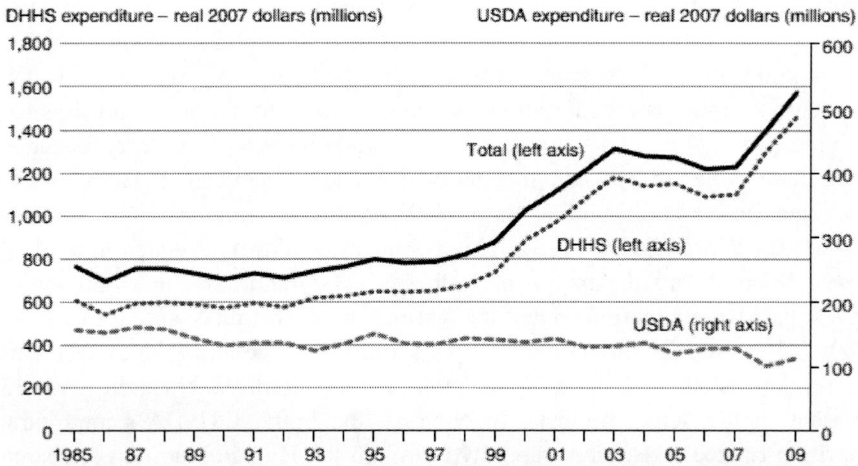

Source: USDA, Economic Research Service based on Human Nutrition Research Information Management data and the Biomedical Research and Development Price Index.

Figure 2. Federal expenditures for nutrition research projects, 1985-2009.

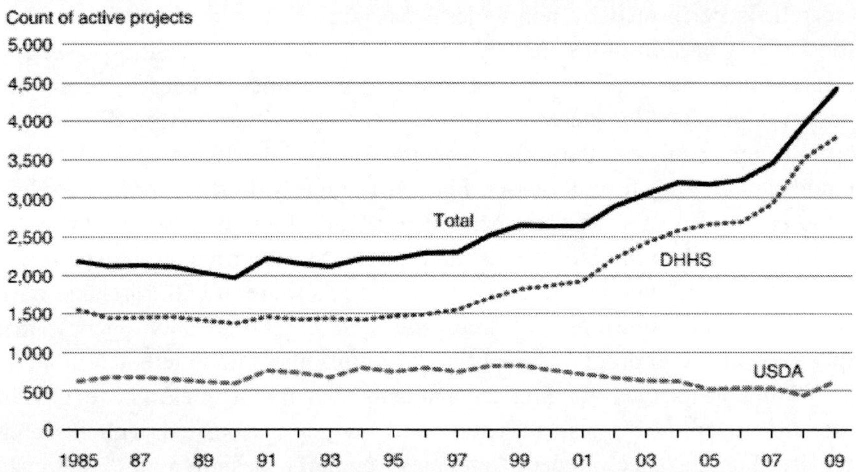

Source: USDA, Economic Research Service based on Human Nutrition Research Information Management data.

Figure 3. Federally supported nutrition research projects, 1985-2009.

The first major distinction within the structure of Federal support is between intramural and extramural research. Intramural research consists of projects initiated and performed by researchers employed by the Federal Government, typically within Government-owned facilities. Maintaining an intramural research program provides the Federal Government with the capacity to anticipate and respond to public needs in a timely and directed fashion. It is often described as using the unique position of a Federal agency to initiate collaborative and translational research intended to facilitate the movement of research findings into practice (http://irp.nih.gov/about-us/what-is-theirp).[9] Within DHHS, the NIH performs most nutrition-related intramural research on its main campus in Bethesda, Maryland. The nutrition topics investigated are determined by the various NIH Institutes and Centers. In 2009, The National Institute of Diabetes and Digestive and Kidney Diseases performed the largest share of nutrition-related intramural research, with 33 percent of the active projects. In contrast, the bulk of USDA's intramural nutrition-related research is supported through the Human Nutrition Research Center Program, which maintains a geographically dispersed network of six nutrition research centers located near research universities. These centers performed over 94 percent of all USDA nutrition-related intramural projects in 2009.

The largest and most diverse category of Federal support, extramural research, is performed by non-Federal organizations using Federal funding. Non-Federal organizations include State governments, universities, colleges, research institutes, hospitals, and private companies. Federal extramural support is obtained by these organizations through either "formula" or "non-formula" mechanisms. Formula mechanisms use algorithms defined in the statutes authorizing formula-based Federal programs. For instance, according to the Hatch Act of 1887 (as amended by Public Law 107-293 in 2002), 52 percent of total authorized funds that go to State agricultural experiment stations are distributed in proportion to the State's share of U.S. rural and farm populations. For nutrition research, the USDA is the only agency that administers Federal programs requiring a formula allocation method.

With non-formula funding mechanisms, Federal agencies have greater discretion over the topics, activities, and budgets associated with research support. For instance, Federal agencies regularly design and disseminate requests for applications (RFAs), which are announcements of research funding opportunities in particular areas of interest. These funds are distributed through contracts, grants, and cooperative agreements. Non-formula funding includes earmarks, agency mission-oriented procurement, as well as competitive (e.g., peer-reviewed) research support.

Maintaining a productive nutrition research system requires understanding how differences between non-formula, formula, and intramural funding mechanisms impact the type of research performed and the rate of scientific discovery. All the mechanisms provide money for research, but the costs and behavioral restrictions they impose on researchers can be quite different. Each mechanism creates incentives that shape the choices and behaviors of researchers. For example, some researchers regard formula funding as predictable and flexible because the statutory rules are stable and the selection of topics is delegated to the researchers' home institution (Huffman and Evenson, 2006). Other researchers favor competitive Federal mechanisms using merit-based, peer-review selection because this system is thought to improve the quality of supported research (PCAST 2012).

A growing academic literature is focused on understanding how research incentives influence the selection of topics, knowledge discovery and dissemination, commercial orientation, and research innovation (Aghion et al., 2005; Azoulay et al., 2011; Dasgupta and David, 1994; Gans and Murray, 2012; Manso, 2011). Some of the incentive design characteristics for promoting innovation and creativity identified in this literature include an emphasis on long-term outcomes with a high tolerance for initial failure, low researcher-time costs for preparing research proposals, few restrictions on how findings are communicated, and flexibility to pursue unanticipated lines of inquiry as research progresses.

From 1985 through 2009, project support through non-formula extramural mechanisms grew by 17 percentage points from 69 to 86 percent (fig. 4). This is the dominant form of support used by DHHS. In the same period, formula-based extramural projects declined from 20 to 8 percent of active projects, and intramural projects fell from 11 to 6 percent. The trends reflect the emergence of DHHS as the principal Federal agency supporting nutrition research, as well as administrative decisions to increase the proportion of non-formula support, particularly competitive grants at USDA. However, the implications of these changes for the nutrition research system are unknown. Collection of new data that link Federal mechanisms to research outcomes would be needed before the effects of these changes on knowledge discovery and dissemination could be assessed.

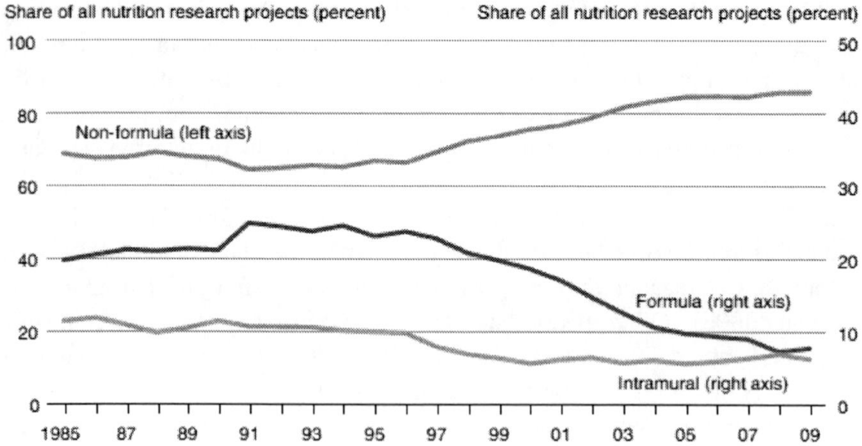

Source: USDA, Economic Research Service based on Human Nutrition Research Information Management data.

Figure 4. Federal support by type of funding mechanism, 1985-2009.

Federal Portfolio of Nutrition Research Areas and Research Performers

From 1985 to 2009, agency budget appropriations largely explained the aggregate trends in Federal support for human nutrition research. These aggregate trends altered the structure of Federal support by changing the funding balance among Federal agencies. To the extent that USDA and DHHS have different missions, use different research funding mechanisms, and support different performers within the nutrition research system, these and other influences may have shifted the distribution of Federal support between nutrition research topics and between the organizations that perform the research.

Defining Portfolio Shares Using HNRIM

The Human Nutrition Research Information Management (HNRIM) data system identifies the nature of the nutrition research conducted within each project. Federal agencies reporting to the system assign 1 or more of 37 nutrition subfield (here, topic area) codes to each project.[10] Using these topic area codes, we defined the portfolio of Federal support as shares (in percentage form) distributed across each nutrition topic in the HNRIM classification scheme. Some projects naturally address multiple topics. For example, in 2004 the USDA supported a project titled "obesity prevention for preschool children." This project was assigned to three topic areas: *infant/child nutrition*, *obesity/anorexia/appetite control*, and *other research in nutrition education*. By assigning individual projects to multiple topic area codes, the HNRIM system captures the practical reality that projects are often relevant to more than one topic area in nutrition research. To construct the portfolio of shares, we parse each project that is assigned to multiple topic areas into the individual components.

We calculated the percentage of projects relevant to each topic by dividing the number of projects assigned to that topic by the total number of project assignments across all 37 topic area codes. Dividing by the total number of assignments from all active projects normalizes the project counts by the level of support. In other words, these shares reflect the distribution of Federal support to each of the 37 nutrition topic areas within a portfolio of fixed size.

Grouping research projects by 37 distinct codes is encyclopedic but cumbersome to present. Distinctions among some codes may be too fine and too numerous for an overview of nutrition support. Fortunately, the HNRIM system provides a means of organizing the codes into six broader research categories:

1. Research on Normal Nutritional Requirements Throughout the Life Cycle

The lifecycle research area includes all projects that have some relevance to the study and understanding of how normal human nutritional needs change for people in different stages of life. This area contains five topic areas: maternal nutrition, infant/child nutrition (0-12 years), adolescent nutrition (13-18 years), adult nutrition (19-65 years), and nutrition of the elderly (65+ years).

Selected project title examples:

- Calcium and Phosphorus Nutrition in High-Risk Infants
- Relation Between Nutritional State and Aging.

2. Research on the Role of Nutrition in Disease, Injury, and Conditions

The disease and conditions area contains all projects that have some relevance to the prevention, amelioration, or treatment of diseases, injuries, infections, or conditions. Medical conditions include all projects that have some relevance for understanding the role of nutrition in genetics, function, and the treatment of conditions such as obesity or anorexia. This area contains 11 topic areas of nutrition research: cardiovascular disease; cancer; trauma; infection and immunology; parenteral/enteral/elemental nutrition; other diseases (e.g., osteoporosis, diabetes); obesity/anorexia/appetite control; genetics and nutrition; nutrition and function; nutrient interactions; and other conditions.

Selected project title examples:

- Modulation of Xenobiotic Toxicity by Diet
- Acetylation and N-Oxidation and Colorectal Cancer.

3. Research on Nutrient Metabolism and Metabolic Mechanisms

The metabolism and metabolic mechanisms area contains all projects that examine certain nutrient variables as well as those projects examining nutrient mechanisms and metabolism not related to research in the areas of life cycle, diseases/injury, and medical conditions given above. This area contains nine topic areas: carbohydrates, lipids, alcohols, proteins/amino acids, vitamins, minerals/essential trace elements, water/electrolytes, fiber, and other nutrients in food.

Selected project title examples:

- Regulation of Albumin Synthesis by Amino Acids
- Physico-Chemical Properties of Carbohydrates in Dairy Foods.

4. Research in Food Sciences

The food sciences research area contains all projects within the various disciplines of food science that have a nutritional component. This area contains four topic areas: food composition R&D, bioavailability of nutrients, effects of technology on foods and diets, and other research in food sciences.

Selected project title examples:

- Improving Physical Properties of Food Proteins by Chemical Modification

- Use of Infrared Heat Processing and Its Impact on Food Quality.

5. Research on Nutrition Monitoring, Education, and Policy

The monitoring, education, and policy area contains all projects that have some relevance to the dietary practices and behaviors of human populations. This area contains six topic areas: nutritional status R&D, food consumption survey R&D, studies of methods for informing and educating the public, other research in nutrition education, studies of dietary practices/ consumption patterns, and the effects of government policy and socioeconomic factors.

Selected project title examples:

- Economic and Behavioral Factors Associated With Food Supplement Usage
- Food Choice Behavior at Point of Purchase: Assessment and Intervention

6. Research on Dietary Supplements

The dietary supplements area contains two topic areas that were added to the HNRIM system in 1999. This area covers all projects that have some relevance for understanding the composition and function of dietary supplements in human nutrition. The two topic areas are: nutrient ingredients of dietary supplements, and botanical and other non-nutrient ingredients in dietary supplements.

Selected project title examples:

- Vitamin B-6 Supplementation and Immune Function
- Multi-vitamins, HAART (highly active antiretroviral therapy) and HIV/AIDS in Uganda.

Distribution of Federal Support Across Major Nutrition Research Areas Is Relatively Stable

Figure 5 displays the Federal portfolio in percentage shares across the six major nutrition research areas for 1985 and 2009. While nearly a quarter of a century separates these two points in time, the distributions across research areas are quite similar. The *Disease, Injury, and Conditions* and *Metabolism and Metabolic Mechanisms* research areas stand out and together constitute 68 and 69 percent of the Federal support in 1985 and 2009, respectively. The stability in these combined shares, however, masks some significant reallocations across the areas. The *Disease, Injury, and Conditions* area grew in portfolio share from 40 percent in 1985 to 49 percent in 2009, while the *Metabolism and Metabolic Mechanisms* area dropped from 28 to 20 percent. The other research area that experienced a significant change in Federal support is *Food Sciences*. The Federal share of active projects in this area declined from 10 to 4 percent. Comparing portfolio shares tells us about how the allocation of support changed over time, but does not account for the overall level of Federal support. In 2009, the Federal Government supported over twice as many active projects as in 1985.

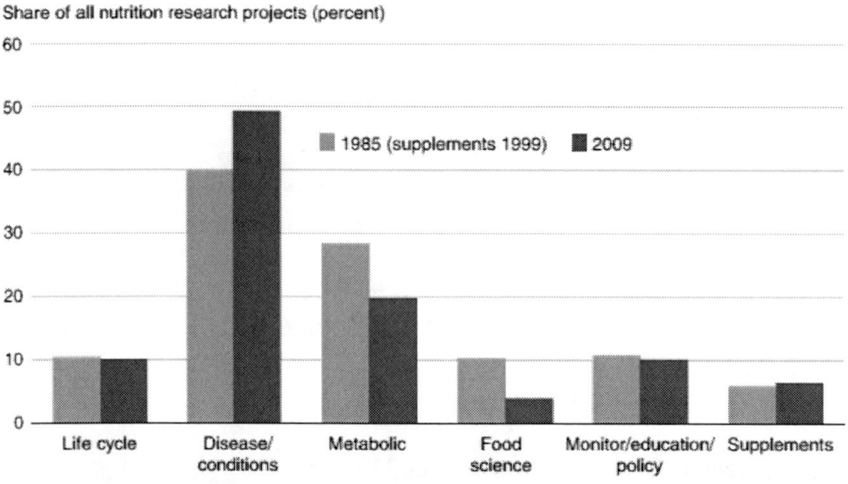

Source: USDA, Economic Research Service based on Human Nutrition Research Information Management data.

Figure 5. Federal nutrition research portfolio: project distributions in 1985 and 2009.

For the *Metabolism and Metabolic Mechanisms* research area, the number of projects supported increased even though the share in the portfolio declined. For *Food Sciences,* both the number and share declined.

Share of Support for *Disease, Injury*, and *Conditions* Research Increased Due to the *Obesity, Genetics*, and Other Diseases Topic Areas

Only 3 of the 11 topic areas composing the *Disease, Injury, and Conditions* research area accounted for most of the growth between 1985 and 2009. The *Obesity/Anorexia/Appetite Control* topic area grew fastest, rising from 3.6 percent to 13.1 percent of the Federal portfolio by 2009 (fig. 6). This topic was the fastest growing within the individual portfolios of both DHHS and USDA. In the DHHS portfolio, the share of active projects related to *Obesity/Anorexia/Appetite Control* increased from 4 percent to 15 percent while the share increased from 1 percent to 7 percent in the USDA nutrition portfolio. The *Genetics* and *Other Diseases* topic areas also grew over this period.[11] The share of active projects related to *Genetics* increased from 3 to 5 percent of the Federal portfolio, while the share related to *Other Diseases* grew from 5 to 9 percent.

Based on the HNRIM data, these increases were not due to changes in the availability of funding through the Federal non-formula, formula, and intramural mechanisms.[12] Instead, program administrators or nutrition researchers determined these trends as a response to influences, such as scientific research opportunities and public needs.

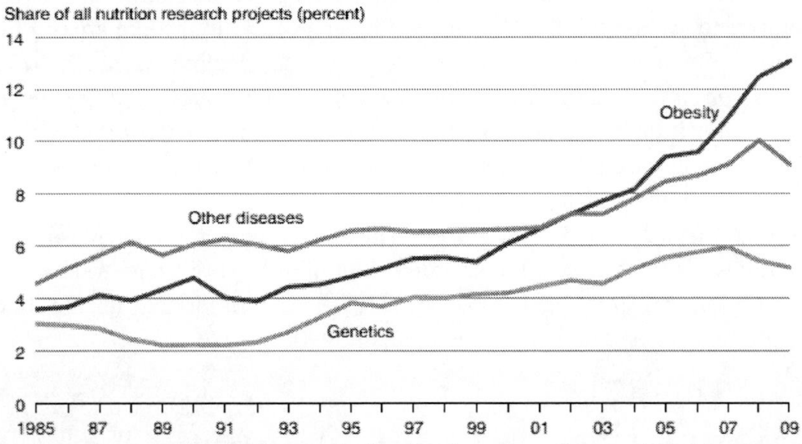

Source: USDA, Economic Research Service based on Human Nutrition Research Information Management data.

Figure 6. Trends in Federal nutrition research portfolio shares for obesity, genetics, and other diseases, 1985-2009.

Share of Support for Metabolism and Metabolic Mechanisms Research Decreased Due to the Vitamins, Minerals, and Proteins Topic Areas

Seven out of the nine topic areas that make up the *Metabolism and Metabolic Mechanisms* research area showed declining shares between 1985 and 2009, with *Vitamins*, *Minerals*, and *Proteins* leading the decline (fig. 7). Between 1998 and 2009, the Federal portfolio share related to *Vitamins* research dropped 2.6 percentage points, from 6.3 to 3.7 percent. *Minerals* and *Proteins* research experienced similar declines. Once again, based on HNRIM data, these decreases were not due to changes in funding availability through the mix of support mechanisms, but they reflect the decisions by program administrators or nutrition researchers.[13]

Share of Support for Food Sciences Decreased Due to the Food Composition and Effects of Technology on Foods and Diets Topic Areas

USDA provides almost all Federal support for nutrition research in the *Food Sciences* area, supporting 80 percent of active projects (on average) from 1985 to 2009. Due to this, USDA support largely determines the overall Federal trend. While the Federal shares for each of the four topic areas within the *Food Sciences* fell, the topics leading the decline were the *Food Composition* and the *Effects of Technology on Foods and Diets*. The Federal share of nutrition research funding going to *Food Composition* research declined steadily from 3.2 percent in 1985 to less than 0.7 percent by 2009 (fig. 8). The Federal share going to research on the *Effects of Technology on Food and Diets* declined from 3.3 percent to 1.2 percent. An increase in the early 1990s seems to reflect the establishment of at least one new food science and nutrition research center by the USDA as authorized in the Food, Agriculture, Conservation, and Trade Act of 1990.

The Portfolio of Federally Supported Organizations That Perform Nutrition Research

Federally supported nutrition research is undertaken by a variety of different organizations. These performing institutions include Federal and

State governments, universities and colleges, private companies, medical schools, hospitals, research organizations, and foundations.

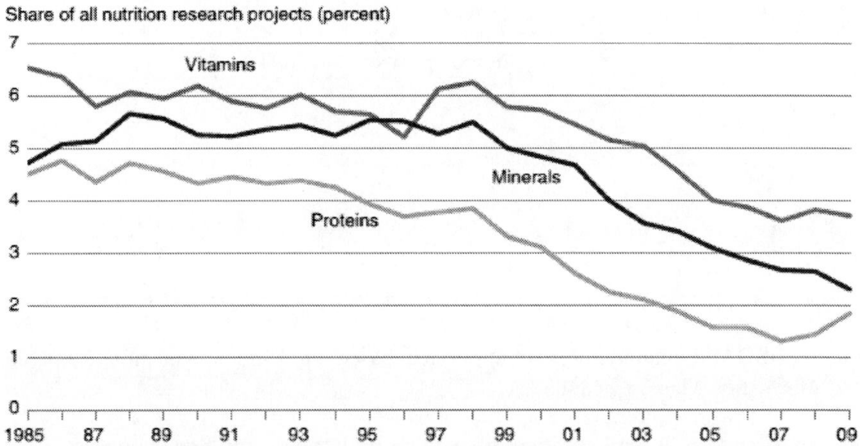

Source: USDA, Economic Research Service based on Human Nutrition Research Information Management data.

Figure 7. Trends in Federal nutrition research portfolio shares for vitamins, minerals, and proteins, 1985-2009.

The distribution of Federal support across these organizations not only reflects the location of skilled research personnel, but also reflects other legal and institutional factors. For instance, the USDA has a unique Federal-State research partnership dating back to the creation of land-grant colleges through the Morrill Act of 1862 and the initiation of formula-based funding for these institutions first authorized by the Hatch Act of 1887. In this partnership, certain universities and colleges are designated by their State legislatures or Congress as cooperating "land-grant" institutions and receive funds on the basis of statutory formulas. Every U.S. State and territory, as well as the District of Columbia, has at least one land-grant institution (APLU, 2012). In addition to these legally based reasons to receive Federal support, performing institutions must also have organizational missions that are consistent with those of the particular Federal agency that provides the funds. So, for instance, nutrition research conducted by medical schools and hospitals is predominantly supported by DHHS and not by USDA.

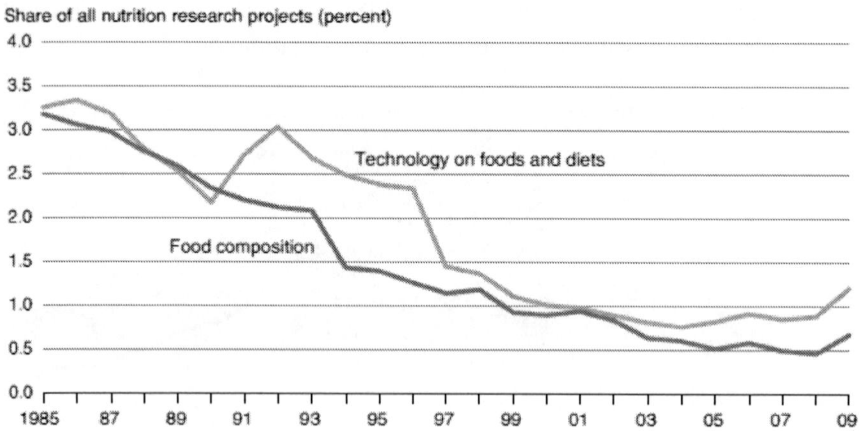

Source: USDA, Economic Research Service based on Human Nutrition Research Information Management data.

Figure 8. Trends in Federal nutrition research portfolio shares for food composition and technology, 1985-2009.

To examine the organizations performing federally supported nutrition research, we classified each organization into one of five groups. The first group, *Government*, includes researchers who work for governmental agencies and consists mostly of research performed in DHHS and USDA facilities supported through intramural funding, but also includes a few State agencies. The second category includes *land-grant universities and colleges* that receive formula funding from the USDA. The data show that nutrition support was widely distributed among these schools. Almost all 1890 land-grant institutions, which are historically Black universities that were established as land-grant institutions under the Second Morrill Act of 1890, received some nutrition support from 1985 to 2009. The third group consists of *non-land-grant universities and colleges*. These research institutions do not receive formula funding from the USDA. Private *companies* that receive Federal support for nutrition research comprise group four. The final group, *Other*, is mostly medical schools and hospitals, but also includes not-for-profit research institutes, foundations, and associations.

Figure 9 illustrates how the portfolio of federally supported organizations that perform nutrition research changed over the 25-year period covered by the HNRIM database. For each group of organizations, two columns are presented. The first column shows the percentage of nutrition projects performed by that group in 1985, and the second column shows the percentage for 2009.[14] The share of Federal nutrition research projects performed by government researchers fell from 12 percent in 1985 to 6 percent in 2009. Land-grant universities are also performing a smaller share of federally supported nutrition projects, a drop of 12 percentage points from 34 percent in 1985 to 22 percent in 2009. The share of federally supported nutrition research performed by the other three groups increased over the 25-year period. Non-land-grant universities and colleges show an increase of 11 percentage points to reach 41 percent of total nutrition projects in 2009. The *Other* category of institutions also increased from 22 percent to 29 percent of all active federally supported nutrition projects. Private companies performed only a small share of projects in both 1985 and 2009.

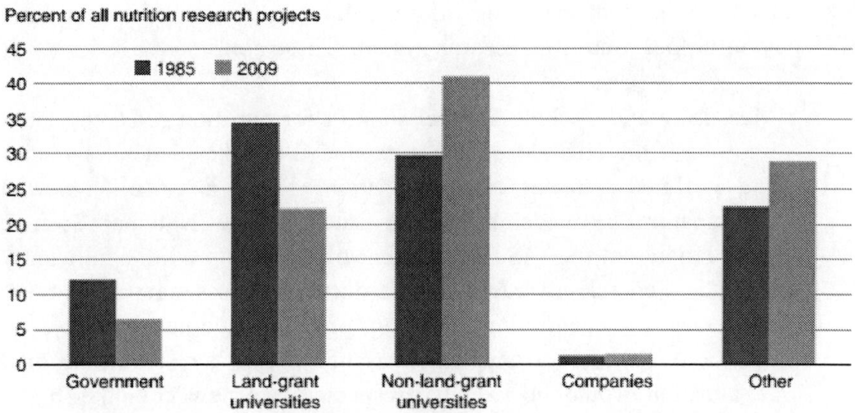

Source: USDA, Economic Research Service based on Human Nutrition Research Information Management data.

Figure 9. Federally supported nutrition research projects by institution type, 1985 and 2009.

CONCLUSION

The U.S. nutrition research system is a vital component of the larger communication network linking nutritional content to dietary choices and health outcomes. It provides the knowledge foundation that fuels continued progress toward better health. Understanding the structure and function of the U.S. nutrition research system, along with the development of diagnostic indicators, can help identify current and emerging issues, such as the impacts of alternative funding mechanisms and policies on research incentives and performance, and research gaps or opportunities requiring further scientific inquiry.

This report takes a first step in that direction, identifying the range of organizations that sponsor nutrition research in the United States and documenting the availability of data on R&D investments. Generally, publicly available data are limited except for Federal investments. The Human Nutrition Research Information Management (HNRIM) database documents the R&D investments by the main Federal agencies supporting nutrition research, DHHS and USDA. The HNRIM data allow us to analyze the level and trends in Federal support across agencies, between topics within the field of nutrition research, and among performing institutions. A number of broad observations emerge that may help inform future studies on the vitality and efficiency of the U.S. nutrition research system. These are:

- Measuring and tracking Federal support for nutrition research would improve if the HNRIM database expanded to other Federal agencies and offered greater standardization. For instance, renewed participation by Federal agencies such as the National Science Foundation and the U.S. Department of Defense would significantly enhance Federal coverage. Standardization across agencies of the processes for identifying nutrition projects, coding the project's nutrition fields, and parsing the share of a project's resources devoted to nutrition would be valuable, especially as new computer-based categorization algorithms like the NIH Research Condition and Disease Classification system are implemented.
- Based on the HNRIM database, this report identifies major shifts in the underlying funding mechanisms used by Federal agencies to support nutrition research. Each of the mechanisms—formula, non-formula, and intramural—use different project selection criteria and impose different costs on individual researchers. Future research examining how alternative Federal mechanisms influence choice of research topic, results, and productivity could guide policy formulation.
- As HNRIM has lost coverage of Federal agencies that support nutrition research, its future utility in helping answer questions about the size and direction of Federal support for nutrition is limited. New initiatives that collect data on private-sector investments in nutrition research would provide useful information.

- The knowledge produced by the U.S. nutrition research system is vital. The volume of knowledge produced and the channels through which this knowledge flows are not well documented. Various commercial databases such as Elsevier's Scopus or Thomson-Reuters Web of Science offer extensive information on professional journal publications that may be used to construct indicators of knowledge output based on publications and knowledge flows based on citation analysis. The patent and trademark data at the U.S. Patent and Trademark Office could also be used to assess nutrition innovation and the commercialization of products and services.

APPENDIX—THE RELATIONSHIP BETWEEN PROJECT COUNTS AND PROJECT AWARDS

Support for nutrition research can be characterized by the number of projects for which the Federal Government provided funding or by the funding itself. If all projects were funded alike—awarded the same dollar amount—it would not make any difference whether support was characterized by number of projects or by dollars. However, when the average financial award per project changes over time, trend analyses using the two approaches will yield different magnitudes, particularly when calculating growth rates.

At the beginning of this study, no project-level financial award amounts were available in HNRIM. In an effort to check the validity of using project counts, the authors evaluated other publicly available databases at the National Institutes of Health (NIH) as potential sources of project-level financial data. The NIH "Awards by Location" Web search page (http://report.nih.gov/award/index. cfm) inventories annual extramural grant and contract awards for research and training projects. Beginning in 1992 (the earliest year), annual project-level data were downloaded, cleaned of obvious errors, and appended to form a project-level database containing project IDs and award amounts. A matching program was used to standardize the project identification numbers for these data. A similar program was written for the HNRIM database. Using these programs, an 88-percent match rate was achieved between the two databases, and the financial award amounts were retrieved and appended to the HNRIM project records for those NIH extramural projects.

Subsequently, the project-level award amounts associated with the HNRIM database became available. These data significantly improved our information on project-level financial awards by adding other Federal agencies such as the USDA, as well as intramural funding amounts. By combining these new data to our matched data, we were able to perform a check on the HNRIM financial amounts for NIH extramural projects. This check verified the accuracy of the HNRIM financial data but also showed that some entries reporting $0 financial awards in HNRIM were not "no-cost" extensions, just missing data. Other than DHHS, all Federal agencies that support nutrition research report $0 as the funded amount for a significant number of projects. Among USDA projects, 15 percent report $0 financial awards. For the other 6 agencies, over 63 percent of the project records have $0 as the funding amount (table A1).

Table A1. Financial information for project-year records, 1985-2009 (%)

Financial data greater than $0	Federal agencies			
	DHHS	USDA	OTHER	Total
Yes	70,062 (92%)	18,327 (85%)	1,153 (37%)	89,542 (89%)
No	5,720 (8%)	3,179 (15%)	1,964 (63%)	10,863 (11%)
Total	75,782	21,506	3,117	100,405

Source: USDA, Economic Research Service calculations based on Human Nutrition Research Information Management and National Institutes of Health databases.

Consistent results across the project count and expenditure approaches depend on the assumption that the average award amount per project remained relatively constant over time. To check this, we calculated the average dollars per project for DHHS and USDA using the HNRIM data. To do so, we imputed missing financial award data. We separated the project-year records with $0 financial awards into two groups: no-cost extensions and missing data. We defined projects as no-cost extensions when they were part of a multiyear project that had at least 1 year of non-zero award information. For these records, the $0 financial award entries in HNRIM are assumed to be correct. For the remainder, we imputed a financial award amount based on the average for the agency and fiscal year. For example, we used the average award amount for non-missing USDA projects in 2009 to replace any missing award data for projects supported by the USDA in 2009.

For DHHS and USDA, figure A1 presents the trends in average award per project in real dollars.[15] In 1985, an average NIH nutrition project received about $393,000 in annual support. This amount peaked at $504,000 in 2001, then declined to $384,000 by 2009. The peak occurred in the middle of the

1999-2003 period over which the NIH research budget doubled (in nominal terms). With this unusual infusion of funds, project size temporarily increased. Comparing project size in 2009 with 1985, it appears that project size declined at an average annual rate of 0.1 percent, but regression analysis showed no statistically significant linear trend in dollars per project over time. Compared to DHHS, USDA project size was more stable. In 1985, USDA nutrition projects received an average of about $250,000. By 2009, average project size had decreased to $180,000, which represents an average annual decrease of 1.4 percent. Again, regression analysis did not show this decline to be statistically different from zero for the 25-year period.

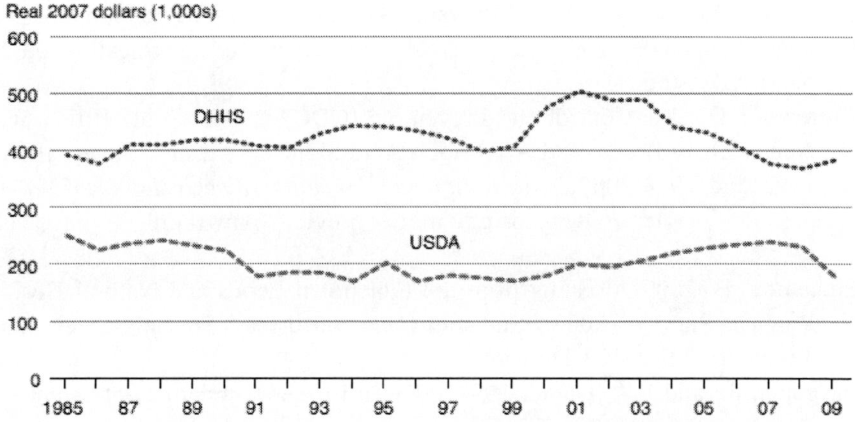

Source: USDA, Economic Research Service based on Human Nutrition Research Information Management data and the Biomedical Research and Development Price Index.

Figure A1. Average dollars per nutrition research project, 1985-2009.

REFERENCES

Aghion, P., M. Dewatripont, and J.C. Stein. 2005. "Acacemic freedom, private-sector focus, and the process of innovation." National Bureau of Economic Research Working Paper No. 11542, Aug.

American Dietetic Association (ADA). 2009. "Position of the American Dietetic Association: Functional Foods," *Journal of the American Dietetic Association* 109(4): 735-746.

Arrow, K. 1962. "Economic Welfare and the Allocation of Resources for Invention," *The Rate and Direction of Inventive Activity* (R.R. Nelson, ed.) Princeton, NJ: Princeton University Press.

Association of Public and Land-grant Universities (APLU). 2012. "The Land-Grant Tradition," http://www.aplu.org/document.doc?id=780.

Azoulay, P., J.S. Graff Zivin, and G. Manso. 2011. "Incentives and creativity: evidence from the academic life sciences," *RAND Journal of Economics* 42(3): 527-554.

Branscomb, L.M., and R. Florida. 1998. "Challenges to Technology Policy in a Changing Economy," *Investing in Innovation* (L.M. Branscomb and J.H. Keller, eds.) Cambridge, MA: MIT Press.

Centers for Disease Control and Prevention (CDC). 2014. "Childhood Obesity Facts" updated Feb. 27, 2014, http://www.cdc.gov/healthyyouth/ obesity/facts.htm

Centers for Disease Control and Prevention (CDC). 2004. "Spina Bifida and Anencephaly Before and After Folic Acid Mandate—United States, 1995-1996 and 1999-2000," *Morbidity and Mortality Weekly Report* 53(17): 362-365, http://www.cdc.gov/mmwr/preview/mmwrhtml/ mm5317 a3.htm.

Cinnamon, B. 2009. "Results from the Functional Foods and Natural Health Products Survey – 2007." Statistics Canada, Business Special Surveys and Technology Statistics Division.

Dasgupta, P., and P.A. David. 1994. "Toward a new economics of science," *Research Policy* 23(5): 487-521.

Davis, C., and E. Saltos. 1999. "Dietary Recommendations and How They Have Changed Over Time." *America's Eating Habits: Changes and Consequences*. U.S. Department of Agriculture, Economic Research Service, AIB-750. http://www.ers.usda.gov/publications/aib750/aib 750. pdf.

Demonty, I., R.T. Ras, H.C.M. van der Knaap, G.S.M.J.E. Duchateau, L. Meijer, P.L. Zock, J.M. Geleijnse, and E.A. Trautwein. 2009. "Continuous Dose-Response Relationship of the LDL-Cholesterol–Lowering Effect of Phytosterol Intake," *The Journal of Nutrition*, 193(2): 271-284.

Finkelstein, Eric. A., Justin G. Trogden, Joel W. Cohen, and William Dietz. 2009. "Annual Medical Spending Attributable to Obesity: Payer and Service-specific Estimates," *Health Affairs* 28(5): w822-w831.

Fuglie, K.O., P.W. Heisey, J.L. King, C.E. Pray, K. Day-Rubenstein, D. Schimmelpfennig, S.L. Wang, and R. Karmarkar-Deshmukh. 2011. *Research Investments and Market Structure in the Food Processing,*

Agricultural Input, and Biofuel Industries Worldwide. U.S. Department of Agriculture, Economic Research Service, ERR-130. Dec.

Gans, J., and F. Murray. 2012. "Funding scientific knowledge: Selection, disclosure and the public-private portfolio," *The Rate and Direction of Inventive Activity: A New Agenda*, Chicago, IL: University of Chicago Press.

General Accounting Office (GAO). 1982. "Progress made in Federal human nutrition research planning and coordination; some improvements needed."

Guenther, P.M., W.Y. Juan, M. Lino, H. Hiza, T. Fungwe, and R. Lucas. 2008. "Diet Quality of Low-Income and Higher Income Americans in 2003-2004 as Measured by the Healthy Eating Index-2005," *Nutrition Insight* 42. U.S. Department of Agriculture, Center for Nutrition Policy and Promotion.

Harvard School of Public Health. 2014. "Genes are not destiny." http://www.hsph.harvard.edu/ obesity-prevention-source/obesity-causes/ genes-and-obesity/#references.

Huffman, W.E., and R.E. Evenson. 2006. "Do formula or competitive grant funds have greater impacts on state agricultural productivity?" *American Journal of Agricultural Economics* 88(4): 783-798.

Interagency Committee on Human Nutrition Research (ICNHR). 1996. "The Human Nutrition Research and Information Management System Definition of Human Nutrition Research: Fifteenth Progress Report." http://hnrim.nih.gov/yearlyreports.aspx.

International Food and Information Council (IFIC). 2013. "Functional Foods Consumer Survey, Executive Research Report." http://www. foodinsight. org/Home.aspx.

Institute of Food Technologists (IFT). 2005. "Functional Foods: Opportunities and Challenges." Institute of Food Technologists Expert Report, http://www.ift.org/Knowledge-Center/Read-IFTPublications/Science-Reports/Scientific-Status-Summaries/Functional-Foods.aspx.

Institute of Medicine (IOM). 2013. "Nutrition education in the K-12 curriculum: The role of national standards: Workshop summary," Washington, DC: The National Academies Press, 2013.

Institute of Medicine. 2005. "Dietary Reference Intakes for Energy, Fat, Fatty Acids, Cholesterol, Protein, and Amino Acids (Macronutrients): A Report of the Panel on Macronutrients, Subcommittees on Upper Reference Levels of Nutrients and Interpretation and Uses of Dietary Reference Intakes, and the Standing Committee on the Scientific Evaluation of

Dietary Reference Intakes." Washington, DC: National Academy of Sciences Press.

Institute of Medicine. 1998. "Scientific opportunities and public needs: Improving priority setting and public input at the National Institutes of Health," Washington, DC: National Academy of Sciences Press, 1998.

Institute of Medicine. 1994. *Opportunities in the nutrition and Food Sciences: Research Challenges and the Next Generation of Investigators*, Washington, DC: National Academy of Sciences Press.

Manso, G. 2011. "Motivating Innovation," *Journal of Finance* 165(5): 1823-1859.

Martinez, Steve W. 2013. *Introduction of New Food Products With Voluntary Health- and Nutrition-Related Claims, 1989-2010,* EIB-108, U.S. Department of Agriculture, Economic Research Service, Feb.

Murphy, K.M., and R.H. Topel. 2006. "The Value of Health and Longevity," *Journal of Political Economy* 114(5): 871-904.

National Research Council (NRC). 2003. "Frontiers in agricultural research: Food, health, environments, and communities," Washington, DC: National Academies of Sciences Press.

National Science Foundation, National Center for Science and Engineering Statistics. 2013. Survey of Research and Development at Universities and Colleges, Integrated Science and Engineering Resources Data System (WebCASPAR) (https://webcaspar.nsf.gov).

Nelson, R.R. 1959. The Simple Economics of Basic Scientific Research," *Journal of Political Economy* 67:297-306.

Ohlhorst, S., R. Russell, D. Bier, D.M. Klurfeld, Z. Li, J.R. Mein, J. Milner, A.C. Ross, P. Stover, and E. Konopka. 2013. "Nutrition Research to Affect Food and a Healthy Life Span," *American Journal of Clinical Nutrition* 98(2): 620-625.

President's Council of Advisors on Science and Technology (PCAST). 2012. *Report to the President on Agricultural Preparedness and the Agricultural Research Enterprise*, Executive office of the President, December 2012.

Price Waterhouse Coopers. 2009. "Leveraging growth in the emerging functional foods industry: Trends and market opportunities."

Rahkovsky, I., S. Martinez, and F. Kuchler. 2012. *New Food Choices Free of Trans Fats Better Align U.S. Diets with Health Recommendations*. EIB-95, U.S. Dept. of Agriculture, Economic Research Service, April.

Rosenberg, I. H. 2009. "History of the USDA Human Nutrition Research Center on Aging at Tufts University," *The Journal of Nutrition* 139: 192-193.

Ribaudo, M., F. Kuchler, and L. Mancino. 2008. "Market Failures: When the Invisible Hand Gets Shaky," *Amber Waves*, U.S. Dept. of Agriculture, Economic Research Service. Nov.

Schimmelpfennig, D., and P. Heisey. 2009. *U.S. Public Agricultural Research: Changes in Funding Sources and Shifts in Emphasis, 1980-2005*. EIB-45, U.S. Dept. of Agriculture, Economic Research Service, March.

University College London. 2013. "How 'obesity gene' triggers weight gain." http://www.ucl.ac.uk/ news/news-articles/0713/15072013-How-obesity-gene-triggers-weight-gain-Batterham.

U.S. Department of Agriculture, Food and Nutrition Service (FNS). 2013. Website for Healthy, Hunger-Free Kids Act of 2010, http://www.fns.usda.gov/cnd/ Governance/Legislation/CNR_2010.htm.

U.S. Department of Health and Human Services, Food and Drug Administration. 2013. "Tentative Determination Regarding Partially Hydrogenated Oils; Request for Comments and for Scientific Data and Information: A Notice by the Food and Drug Administration on 11/08/2013. *Federal Register* Vol. 78, No. 217: 67169-67175. http://www.gpo.gov/fdsys/pkg/FR-2013-11-08/pdf/2013-26854.pdf.

Viscusi, W.K., and J.E. Aldy. 2003. "The value of a statistical life: A critical review of market estimates throughout the world," *The Journal of Risk and Uncertainty* 27(1): 5-76.

End Notes

[1] The 1995 Dietary Guidelines was the first mandated joint report, prepared according to the 1990 National Nutrition Monitoring and Related Research Act. Among the requirements, reports are reviewed by a committee of experts, updated if necessary, and published every 5 years.

[2] National Nutrition Monitoring and Related Research Act.

[3] There are other inputs into determining food benefits for WIC as the Guidelines do not cover children under age 2, a major part of the WIC caseload.

[4] We could not find any data sources that report human nutrition research by State governments.

[5] In this report all data are presented in fiscal years (FY) and are referred to simply as years. The most recent data reported are for FY 2009 (October 1, 2008 – September 30, 2009). The method for assigning projects to the field of nutrition and its subfields (or topic areas) changed in ways that make it impossible to compare pre- and post-2009 data.

[6] Throughout this report, the annual number of active project records is simply referred to as projects, even though these may be ongoing components of broader multiyear research efforts.

[7] This report does not count Federal support for company innovation or commercialization activities, facilities construction, or repair as part of research or training. Research and training projects are not analyzed separately and the combination is simply referred to as nutrition research.

[8] The Federal roles of the USDA and DHHS in human nutrition research were established in the National Agricultural Research, Extension, and Teaching Policy Act of 1977. That legislation directed the USDA to establish a separate and distinct mission to support research into food and human nutrition. It also designated the USDA as lead agency of the Federal Government for human nutrition research except for the biomedical aspects that relate to the diagnosis and treatment of disease, which were reserved for the U.S. Department of Health, Education, and Welfare (HEW), which later became DHHS. Prior to the passage of the Act, the Secretaries of HEW and USDA recognized their mutual interests in human nutrition research and signed an agreement for sharing their responsibilities (Rosenberg, 2009).

[9] Translational research is a broad category of research activities that are intended to facilitate the application of basic research and thereby improve health and well-being. The National Cancer Institute, Translational Research Working Group, provides a more detailed definition at www.cancer.gov/researchandfunding/trwg/TRWG-definition-and-TRcontinuum.

[10] In 1980, the Joint Subcommittee on Human Nutrition Research developed a 34-code classification system for human nutrition research that was expanded to 35 codes in 1985. Two codes were added in 1999 for research relevant to nutrition supplements, and the HNRIM database also includes 6 codes for areas of special interest to the NIH. Our analysis is based on the 37-code system used by both USDA and DHHS (ICHNR, 1996).

[11] The Other Diseases topic area includes nutrition research for all diseases except those related to cardiovascular and cancer. For instance, this topic area includes diseases related to eyes, liver, kidney, lung, renal, and mental health.

[12] The Federal share for these topics increased across all the funding mechanisms despite the budget reductions for formula and intramural support. This fact rules out funding availability as the driving reason for the observed increases.

[13] The Federal share for these topics decreased across all the funding mechanisms despite the budget increases for non-formula support. This fact rules out funding availability as the driving reason for the observed decreases. We investigated the possibility that the introduction of the dietary supplements research area may have caused a reclassification of projects related to vitamins, minerals, and proteins that accelerated the decline in this topic area after 1999. The data do not suggest reclassification is the main driver, but it may have contributed.

[14] Due to the large number of performing institutions that required manual searches to determine the proper classification, we only analyzed organizations active in 1985 and 2009.

[15] Adjustments for inflation were made using the Biomedical Research and Development Price Index, 2007 reference year. The index is available at http://officeofbudget.od.nih.gov/gbiPriceIndexes.html.

In: U.S. Nutrition and Agricultural Research ISBN: 978-1-63484-170-2
Editor: Arlene Ramos

Chapter 2

AGRICULTURAL RESEARCH: BACKGROUND AND ISSUES*

Jim Monke

SUMMARY

The U.S. Department of Agriculture's (USDA's) Research, Education, and Economics (REE) mission area has the primary federal responsibility of advancing scientific knowledge for agriculture through research, education, and extension. USDA REE responsibilities are carried out by four agencies: the Agricultural Research Service (ARS), the National Institute of Food and Agriculture (NIFA), the Economic Research Service (ERS), and the National Agricultural Statistics Service (NASS). USDA conducts its own research and administers extramural federal funding to states and local partners primarily through formula funds and competitive grants.

Discretionary funding for the REE mission area totaled $2.726 billion in FY2015, and mandatory funding from the 2014 farm bill adds another $114 million per year on average.

Debates over the direction of public agricultural research and the nature of its funding mechanism continue. Ongoing issues include the need, if any, for new federal funding to support agricultural research, education, and extension

---

* This is an edited, reformatted and augmented version of a Congressional Research Service publication R40819, prepared for Members and Committees of Congress, dated July 13, 2015.

activities, and the implications of allocating federal funds via formula funds versus competitive grants. Many groups believe that Congress needs to increase support of U.S. agriculture through expanded federal support of research, education, and extension programs, whereas others believe that the private sector, not taxpayer dollars, should be used to support these activities.

USDA's Research, Education, and Economics (REE) Mission Area

The U.S. Department of Agriculture (USDA) combines its research activities into the Research, Education, and Economics (REE) mission area. The mission area is composed of four agencies with the federal responsibility to advance scientific knowledge for agriculture. Activities include the biological, physical, and social sciences related broadly to agriculture, food, and natural resources, delivered through research, statistics, extension, and higher education (*Table 1*).

Table 1. USDA's Research, Education, and Economics (REE) Mission Area
(FY2015 discretionary budget authority)

Function	Entity	Description
	Agricultural Research Service (ARS) $1.133 billion salaries and expenses + $45 million for buildings & facilities	ARS conducts research and disseminates information related to crop and livestock production and protection, human nutrition, food safety, rural development, natural resource management, and conservation. Emphasis is on national and regional problems, including higher-risk and long-term research such as plant and animal genome programs. Workforce of about 7,400

Function	Entity	Description
		full-time employees located across about 100 research stations.
Intramural Federal research	National Agricultural Statistics Service (NASS) $172 million	NASS collects and compiles statistics related to U.S. agriculture (e.g., Census of Agriculture, crop forecasts, estimates of farm prices). Workforce of about 1,100 fulltime employees; offices in DC, 45 states, and Puerto Rico.
	Economic Research Service (ERS) $85 million	ERS provides economic and policy analysis to inform public and private decision making related to food, farming, natural resource management, agricultural markets, and rural development. Workforce of about 370 full-time employees entirely located in Washington, DC.
Extramural Federal funding of state and other institutions	National Institute of Food and Agriculture (NIFA) $1.290 billion	NIFA leads and funds external research, extension, and educational programs for agriculture, the environment, human health and well-being, and communities. Provides grants and partnerships with the land-grant university system and other organizations that work at the state and local level. Workforce of about 410 employees. Federal

Table 1. (Continued)

Function	Entity	Description
		funding through competitive grants and "formula funds" (the latter based on each state's farm and rural population, with matching fund requirements from the states).
Administrative	Under Secretary for Research, Education and Economics Chief Scientist Research, Education, and Extension Office (REEO) $898,000	The Under Secretary is designated as the Chief Scientist and leads the REEO, which coordinates USDA programs, sets priorities, and aligns scientific capacity across the four agencies. Six divisions: (1) renewable energy, natural resources, and environment; (2) food safety, nutrition, and health; (3) plant health and production; (4) animal health and production; (5) agricultural systems and technology; and (6) agricultural economics and rural communities.
REE total	$2.726 billion	

Source: CRS, using USDA and appropriations committee information. Amounts are FY2015 budget authority. For program details, see USDA's Congressional Budget Justification at http://www.obpa.usda.gov.

Although all four USDA research agencies are headquartered in Washington, DC, much of the work is executed through a set of agency field stations and a network of university partners throughout the United States that operate at the state and local levels. The Economic Research Service (ERS) is the only REE agency entirely based in Washington, DC. The Agricultural

Research Service (ARS) has about 100 research centers and work locations across the United States, Puerto Rico, and the Virgin Islands (*Figure 1*). The National Institute of Food and Agriculture (NIFA) partners with colleges of agriculture at land-grant universities in 50 states and eight U.S. territories, affiliated state agricultural experiment stations (SAESs), schools of forestry and veterinary medicine, and the Cooperative Extension system.[1] Besides the traditional "1862 institutions," or land-grant colleges of agriculture established by the Morrill Act of 1862, there are 19 historically black colleges and universities of agriculture ("1890 institutions;" HBCUs) created by the Second Morrill Act of 1890, and 31 Native American colleges (referred to as tribal colleges) that gained land-grant status in 1994 (*Figure 2*).

Federally funded intramural research is intended in part to address issues of national importance and to promote basic research, regional coordination, and spillover.[2] Federally funded extramural research activities are decentralized and often are regionally specific and/or applied in nature. The federal-state research system also supports USDA's regulatory programs in the areas of meat, poultry, and egg inspection; foreign pest and disease exclusion; and control and eradication of crop and livestock threats; among other things.

The House and Senate Agriculture committees are responsible for the authorizing statutes and oversight of agricultural research, education, and extension programs. Congress traditionally considers research, education, and extension policy as one title of periodic omnibus farm bills, most recently in 2014 (P.L. 113-79, Title VII). Annual appropriations bills and hearings provide more frequent opportunities for oversight and determination of funding.

Federal Funding

The majority of federal funding for agricultural research, education, and extension activities is from annual discretionary appropriations in the Agriculture and Related Agencies appropriations bill. In FY2015, discretionary funding for the entire REE mission area totaled $2.726 billion (*Table 1*). A subset of research programs, especially within NIFA, is provided with mandatory funding—such as for specialty crops or organic agriculture that were created in the 2008 and 2014 farm bills. The 2014 farm bill provides an average of $114 million per year of mandatory funding to agricultural research (*Figure 3*).[3]

Intramural federal agencies (ARS, NASS, and ERS) are funded directly with discretionary appropriations to pay salaries and expenses of federal employees, to conduct research, and to build and maintain facilities.

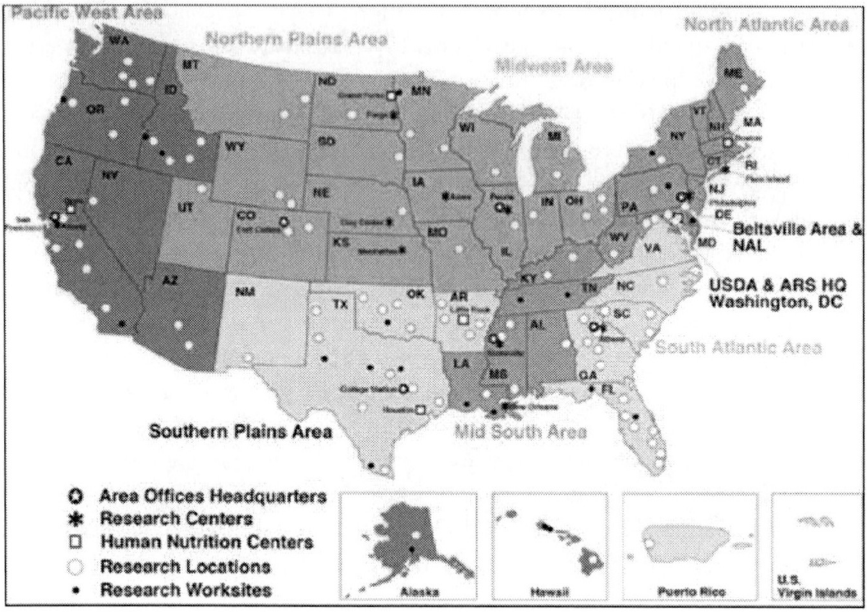

Source: USDA, Agricultural Research Service, http://www.ars.usda.gov/pandp/locations.htm.

Figure 1. USDA Agricultural Research Service Locations.

The agency for extramural research (NIFA) retains a small part of its appropriation for salaries and expenses of a relatively small cadre of employees to administer its programs. The vast majority of the NIFA appropriation is available for extramural research grants that are made primarily through two types of funding (*Figure 3*).

1. *Formula Funds* for research and extension are awarded to land-grant institutions (1862, 1890, and 1994 institutions), schools of forestry, and schools of veterinary medicine following distribution formulas that are set in statute. The amount of funds provided to each institution is determined by census-based statutory formulas that change infrequently. Research priorities are more geographically focused as local or regional university leaders decide which specific

projects will be supported by an institution's formula grant allotment (7 U.S.C. 301 et seq.). Two accounts provide most of the formula funding.[4]

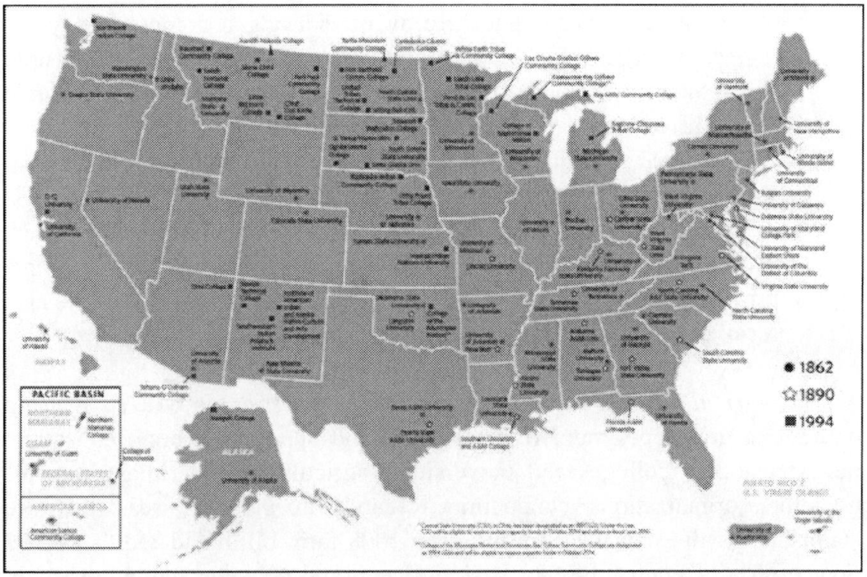

Source: U.S. Department of Agriculture, National Institute of Food and Agriculture, http://nifa.usda.gov/sites/default/files/resource/lgu_map_6_25_2014_0.pdf.

Figure 2. Land-Grant Colleges of Agriculture.

- The *Hatch Act* of 1887 authorizes research funding at the *state agricultural experiment stations* (SAESs). Congress amended the Hatch Act in 1955 to include a formula that distributes the federal appropriation among states based on each state's farm and rural population as determined by the U.S. Census (7 U.S.C. 301). The Hatch Act, as amended, also requires dollar-for-dollar matching funds from state appropriations, but most states appropriate three to four times the federal allotment.[5] The act also requires each state to use 25% of its Hatch Act funds to support multi-state or regional research. HBCUs get similar funding through *Evans-Allen* research grants and *1890s capacity building grants*.
- The *Smith-Lever Act* provides federal formula funds for *cooperative extension* activities using statutory formulas and non-federal matching requirements similar to the Hatch Act (7 U.S.C. 341).

Federal funding supporting forestry and veterinary programs at the land grant institutions also is distributed among the institutions according to formulas, but these have different criteria than the Hatch Act and Smith-Lever Act formulas.

2. Competitive Grants. Funds are awarded using a competitive, peer-reviewed process for fundamental and applied research, extension, and higher education activities, as well as for projects that integrate research, education, and extension functions. Competitive programs are designed to enable USDA to attract a wide pool of applicants to work on agricultural issues of national interest and to select the best quality proposals submitted by highly qualified individuals, institutions, or organizations (7 U.S.C. 450i(b)). Competitive grants primarily are funded with discretionary appropriations, but some also receive mandatory funding from the farm bill.

The *Agriculture and Food Research Initiative (AFRI)* is NIFA's flagship competitive grants program. It funds basic and applied research, education, and extension to colleges and universities, agricultural experiment stations, and other organizations conducting research in priority areas that are established partially in the farm bill. The 2008 farm bill mandated that AFRI allocate 60% of grant funds for basic research and 40% for applied research. At least 30% of total funds must be used to integrate research with education and/or extension activities.

Formula Funds vs. Competitive Grants

Much debate has surrounded the implications of various funding mechanisms for agricultural research, in particular federal formula funding vs. external peer-reviewed competitive grant funding. Efforts to improve public agricultural research efficiency have often included calls to increase the use of competitive grants as a means of more effectively allocating limited federal resources. Two historically influential reports published by the National Academy of Sciences (NAS)[6] and the Rockefeller Foundation[7] argued that the agricultural research of 30-40 years ago had become overly focused on applied research rather than cutting-edge basic research, and both reports recommended a shift in funding mechanisms toward more use of competitive funding rather than formula funding of the state agricultural experiment stations (SAESs).

In its 2012 report, the President's Council of Advisors on Science and Technology recommended a continued focus on increasing the proportion of research funds award competitively.[8]

The creation of a new, separate grant-making agency within USDA that was solely responsible for administering competitive grants programs in agricultural research and extension was one of the recommendations that came out of a National Academy of Sciences 2000 report looking at the efficacy of the National Research Initiative Competitive Grants Program.[9] In July 2004, the USDA task force published its report, which advocated for the concept of a new NIFA.[10] The task force modeled NIFA on the National Institutes of Health (NIH) and National Science Foundation (NSF), recommending that NIFA should accomplish its mission primarily through administering competitive peer-reviewed grants that support and promote high-caliber, fundamental agricultural research.

The choice of funding mechanisms is viewed by some as important because it is thought to determine where and by whom the research is conducted, and the type of research performed. On the one hand, the competitive, peer-reviewed process is thought to have an advantage because a wider pool of candidates is eligible to apply for funding (e.g., grant recipients are not limited to land-grant institutions or SAESs), and it is thought to engage the best and brightest minds in addressing challenges facing the agriculture sector.

At the same time, the agricultural community widely acknowledges that USDA-funded research has an important role to play, whether carried out intramurally or through formula funds. Using a census-based formula to calculate the annual distribution of research funds to each state has meant that the state allocations have been quite constant from year to year, since annual appropriations have remained relatively level. Although all federal sources account for 30% or less of total funding for the experiment stations (including grants from non-research agencies within USDA and from other federal departments), the reliability of the formula funds has resulted in them traditionally being used to support the core ongoing research programs of the state agricultural experiment stations, which underpin academic programs at many universities.

Studies have shown that funding through competitive grants tends to favor basic research, reach a greater proportion of non-land grant universities, and are concentrated among fewer states than funding that is allocated by statutory formula funds. States with large agricultural production and top-ranked academic programs in biology and agricultural sciences were generally more competitive and more successful in receiving larger shares of federal funds allocated as competitive grants.[11]

Other studies have shown that federal formula funding has a larger impact on agricultural productivity over the longer term than federal competitive grants and contracts.[12] The rationale is that federal-level research is steady funding that can support core or foundation research and is best able to take on higher-risk and long-term projects of national importance, such as deciphering plant and animal genomes, conducting longitudinal studies on human nutrition, and measuring and analyzing current and historical socioeconomic factors in the U.S. food and fiber sector. Proposals that address problems of concern to an entire state or region, and/or are multidisciplinary, are typically underfunded in a national competitive-grant process, despite the fact that such research problems are considered by many to be of critical concern and may have a large net social payoff to the agricultural sector.

Intramural vs. Extramural Funding

ARS is the principal in-house or intramurally funded research arm of the USDA. Many believe that maintaining some level of federally funded internal research allows ARS to fill an important niche, not met by industry or other institutions, specifically to address research problems of national and long-term priority, such as conservation and improvement of plant genetic resources, surveillance and monitoring of national and regional disease outbreaks, soil and water resource management, and adaptation to increasing climate variability and extreme events. On the other hand, some believe that ARS scientists have an unfair advantage in competing with other agricultural scientists, who do not have an endowed source of support for core research expenditures.

PUBLIC AND PRIVATE FUNDING

A recurring policy issue is whether or not there is a need for more federal spending for agricultural research, education, and extension. A related issue is the role of publicly funded research in context with privately funded research.

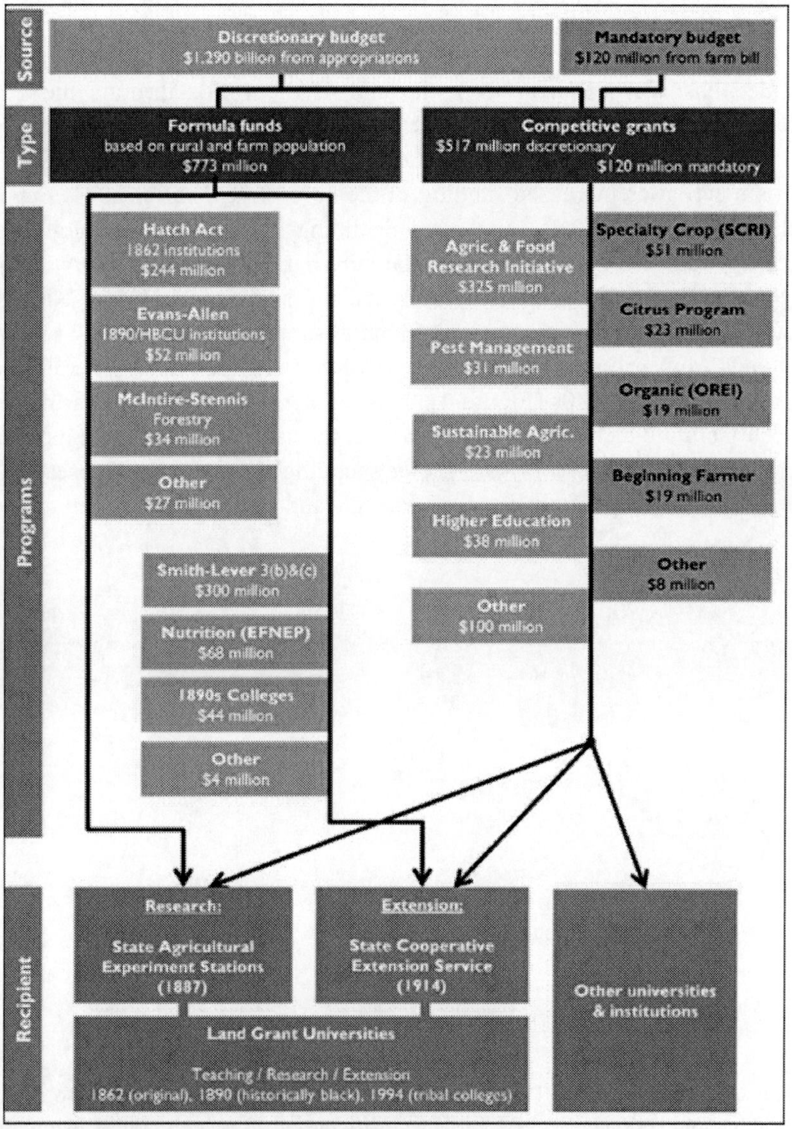

Source: CRS, based on USDA FY2016 Budget Summary.

Notes: AFRI=Agriculture and Food Research Initiative; EFNEP=Expanded Food and Nutrition Education Program; OREI=Organic Agriculture Research and Education Initiative; SCRI=Specialty Crop Research Initiative.

Figure 3. National Institute of Food and Agriculture (NIFA) Budget. (FY2015 budget authority).

Over the long run, and adjusted for inflation, public funding for agricultural research grew steadily from the 1950s to the late 1970s, and remained basically constant from the end of the 1970s through the 1980s (Figure 4). There was a marked rise in public funding from 1998 through 2001, at a time of a budget surplus. One-time, supplemental funding for anti-terrorism activities increased funding in the several years after 2001. Funding levels peaked in FY2010, but began declining in FY2011 as Congress cut federal spending.13 As a result of a relatively flat or declining USDA research budget, funding from other federal agencies, such as NIH and NSF, has accounted for an increasing portion of federal support for agricultural research.

Funds from private industry for agricultural research also generally have increased since the 1970s (Figure 4). This includes public-private partnerships that can facilitate technology transfer and at the same time help to supplement federal and state support. Private-sector spending on agricultural research has grown faster than publicly funded research and development over the long term.

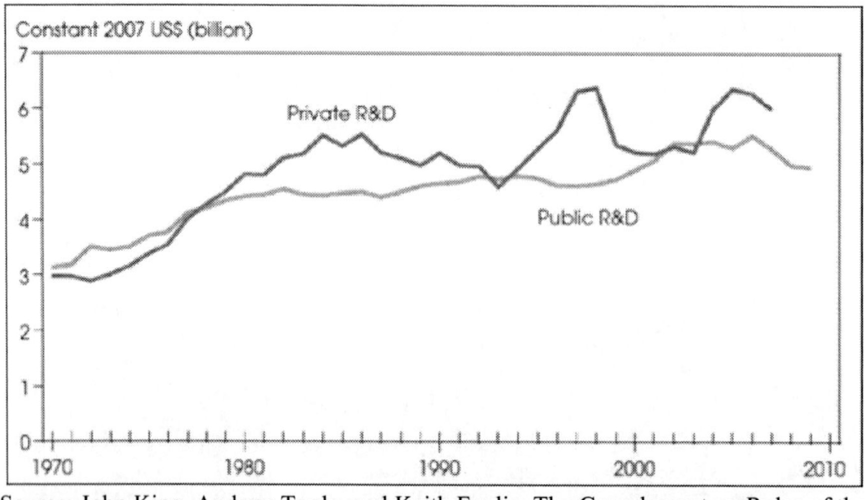

Source: John King, Andrew Toole, and Keith Fuglie, The Complementary Roles of the Public and Private Sectors in U.S. Agricultural Research and Development, USDA Economic Research Service, Economic Brief No. (EB-19), September 2012, at http://www.ers.usda.gov/media/913804/eb19.pdf.

Figure 4. Real U.S. Public and Private Agricultural R&D Expenditures.

Figure 5 shows the many funders of agricultural research, the scale of their contributions, and the destinations for that funding. In 2009, out of $13.9 billion of agricultural research funding, about 68% came from the private sector ($9.5 billion) and about 62% remained in the private sector ($8.7 billion). State governments passed through $1.4 billion to land-grant universities (LGUs) or state agricultural experiment stations (SAESs), which incidentally received a majority of their funding from combined federal sources in nearly equal shares from USDA, other federal agencies such as NSF and NIH, and the private sector.[14] USDA intramural research by ARS, ERS, and NASS accounted for about 11% of total agricultural research spending in 2009.

Some observers are concerned that both the increase in non-USDA public funding and the increase in private funding might cause the focus of agricultural research to shift away from the U.S. agricultural sector's highest priorities and needs. They believe that such a shift could hamper the nation's ability to remain cutting-edge with regard to new innovations, to be competitive in a global market, and to cope with long-term challenges such as pest and disease outbreaks, climate change, and natural resource management.

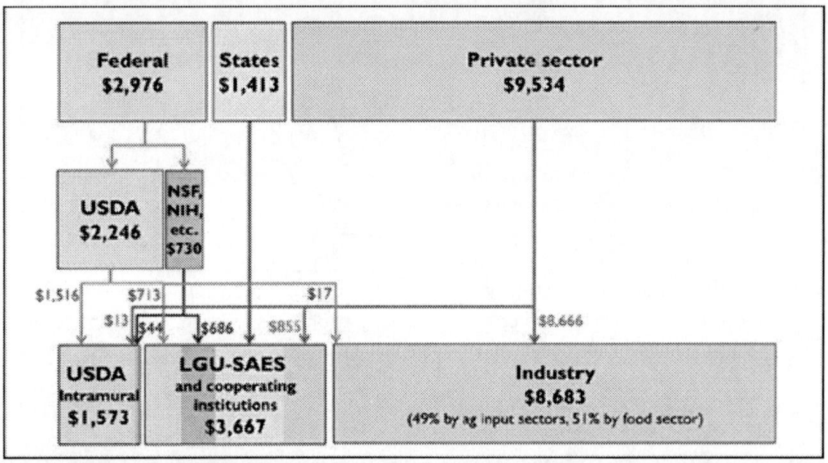

Source: USDA, Economic Research Service (update by CRS of Figure 2 in USDA-ERS EB-19, September 2012, based on ERS data obtained by personal communication in June 2015).

Notes: Includes research and development funding only; that is, does not include Cooperative Extension.

Figure 5. Funders and Performers of U.S. Food and Agricultural Research in 2009. (dollars in millions).

AGRICULTURAL RESEARCH SUPPORTS PRODUCTIVITY

Public investment in agricultural research has been linked to productivity gains and economic growth.[15] Studies consistently report high social rates of return (20%-60% annually) from public agricultural research.[16] The rate of return may depend on the type of research (basic vs. applied), the duration of the research investment, and the specific commodity being studied.

Advances in agricultural research and extension were critical to the huge productivity gains in the United States after World War II (*Figure 6*).[17] Yields for some major crops grew about 2% annually from the 1950s through the 1980s, but that growth has moderated since 1990 (*Table 2*).

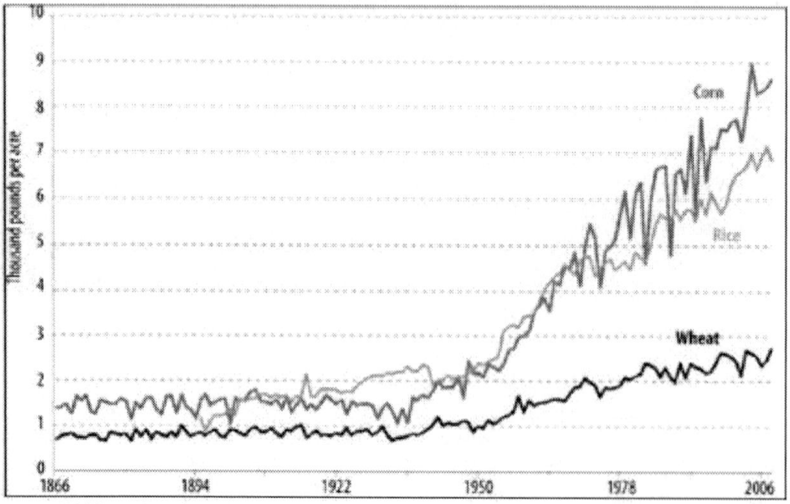

Source: Philip Pardey, "Putting U.S. Agricultural R&D and Productivity Developments in Perspective," Farm Foundation Conference, 2009, http://www.farmfoundation.org/news/articlefiles/1705-Pardey%20.pdf.

Figure 6. U.S. Farm Commodity Yields, 1866-2008.

Table 2. Percentage Growth Rates per Year of Average U.S. Crop Yields

	Corn	Wheat	Rice	Soybeans
1950-1989	2.85	1.75	2.27	1.02
Post-1990	1.5	0.15	1.37	1.16

Source: J. M. Alston, J. M. Beddow, and P. G. Pardey, "Agricultural Research, Productivity, and Food Commodity Prices," University of California, 2009, http://giannini.ucop.edu/media/are-update/files/articles/ v12n2_5.pdf.

Advances in the basic and applied agricultural sciences, such as disease-resistant crop varieties, efficient irrigation practices, and improved marketing systems, are considered fundamental to achievements in agricultural yields, increases in farm sector profitability, higher competitiveness in international agricultural trade, and improvements in nutrition and human health.

Some want more public spending on agricultural research to maintain U.S. competitiveness and to increase agricultural productivity in the face of world population growth and food demand.[18] But agricultural research competes for federal funding in relation to other federal agricultural programs, such as conservation, farm income and risk management programs, food safety inspection, rural development, and domestic and foreign food aid programs.[19]

The 2008 farm bill required the REEO to develop and implement a USDA Roadmap for Agricultural Research, Education, and Extension to plan and coordinate across the entire department both capacity and competitive programs, as well as USDA-administered intramural and extramural programs.[20] The objective was to identify current trends, constraints, gaps, and major opportunities that no single entity within the USDA would be able to address individually. The research provisions, including changes to the management and structure of REE in the 2008 farm bill, drew heavily on proposals and recommendations put forth by key stakeholder groups, including the USDA Task Force on Research, Education, and Extension and the Association of Public and Land-Grant Universities (APLU).[21] USDA subsequently published an "Action Plan" that builds upon the Roadmap.[22]

Some argue that the stagnant growth in inflation-adjusted USDA funding for agricultural research, education, and extension over the past few decades has hurt the ability of the U.S. agricultural sector to stay productive and competitive.[23] It is widely acknowledged that new innovations and technologies related to production, processing, marketing, and natural resource management are essential for continued productivity gains and economic growth of the sector.

Some of these same critics argue that USDA has not been successful at elevating agricultural research to the same priority level with policymakers as other sectors, such as health, and that U.S. agriculture will suffer over the long term because of a lack of new innovations. These critics argue that the lack of public investment in new agricultural innovations will have dire consequences in the future, especially given new and varied challenges, such as rising production costs, especially for fuel and inputs; new pest and disease outbreaks; increasing frequency of extreme weather events, such as droughts and floods; and climate change.

On the other hand, some argue that the federal government should have a limited role in funding agricultural research and that taxpayer dollars should not be used to support what should be a private sector endeavor. In addition, due to a severely constrained federal budget in recent years, limited resources are available to support the agricultural sector. Historically, Congress has not prioritized increasing funding for agricultural research, education, and extension activities, and instead has tended to fund programs designed to provide more immediate benefits to farmers, such as income support and crop insurance. Others believe that the states and the private sector should fill the research funding gap left by the federal government.

At the same time, while private sector funding has increased over time to fill some of the gap in public spending, there is growing concern that private sector funding focuses primarily on taking existing technologies to market (i.e., more applied research) and does not focus on basic problems and/or longer-term challenges that the agricultural sector may face in the future, such as environmental sustainability or adaptation to climate variability.

Some research advocates have argued that some of USDA's current agricultural research portfolio duplicates private sector activities on major crops, including corn, soybeans, wheat, and cotton.[24] They argue that funding should be reallocated to basic, noncommercial research to benefit the public good that is typically not addressed through private efforts. Others point out that these crops are economically important to the food, feed, and energy sectors and should continue to receive significant amounts of public funding, especially for emerging threats, such as new pests and pathogens, limited water availability, and impacts of agriculture on human and environmental health.

End Notes

[1] The Cooperative Extension system is a nationwide, non-credit educational network where each U.S. state and territory has a state office at its land-grant university and a network of local or regional offices. The purpose of Extension is to gather knowledge gained through research and education and deliver it for practical use directly to farmers and other residents (rural and urban).

[2] Economists use the term "spillover" to capture the idea that some of the economic benefits of research and development activities affect agents, locales, or activities beyond the original research purpose or location.

[3] CRS Report R42484, Budget Issues That Shaped the 2014 Farm Bill.

[4] Additional details are available in the USDA Budget Summary and Explanatory Notes for NIFA, available at http://www.obpa.usda.gov.

[5] An exception exists in statute for states that cannot meet the matching requirement. The lack of matching funding for some institutions, however, reduces the resources available to often minority-serving institutions (see Association of Public and Land-Grant Universities, Land-Grant But Unequal: State One-to-One Match Funding for 1890 Land-Grant Universities, September 2013, at http://www.aplu.org/library/land-grant-but-unequal-state-one-to-one-match-fundingfor-1890-land-grant-universities/file.

[6] National Academy of Sciences (NAS), Report of the Committee on Research Advisory to the U.S. Department of Agriculture, National Academy Press, Washington, DC, 1972.

[7] Rockefeller Foundation, Science for Agriculture, The Rockefeller Foundation, New York, NY, 1982.

[8] President's Council of Advisors on Science and Technology, Report to the President on Agricultural Preparedness and the Agriculture Research Enterprise, Washington, DC, December 2012, p. 31, http://www.whitehouse.gov/sites/ default/files/microsites/ostp/pcast_agriculture_20121207.pdf.

[9] National Academy of Sciences, National Research Initiative: A Vital Competitive Grants Program in Food, Fiber, and Natural-Resources Research, Washington, DC, 2000, http://www.nap.edu/catalog/9844.html.

[10] National Institute for Food and Agriculture: A Proposal, report of the Research, Education, and Economics Task Force of USDA, July 2004. The report is available at http://www.ars.usda.gov/sp2userfiles/place/00000000/ national.doc.

[11] Kelly Day Rubenstein, Paul W Heisey, and Cassandra Klotz-Ingram et al., "Competitive Grants and the Funding of Agricultural Research in the United States," Review of Agricultural Economics, vol. 25, no. 2 (September 24, 2003), pp. 352-368.

[12] Wallace E. Huffman and Robert E. Evenson, "Do Formula or Competitive Grant Funds Have Greater Impact on State Agricultural Productivity," American Journal of Economics, vol. 88, no. 4 (November 2006), pp. 783-798.

[13] While REE funding reductions since FY2010 have contributed to deficit reduction, critics are concerned with impacts on research programs. For example, universities reported widespread delays and reductions in research activities as a result of sequestration (Association of Public and Land-Grant Universities, Survey on Sequestration Effects—Selected Results from Private and Public Research Universities, Washington, DC, November 11, 2013, http://www.aau.edu/WorkArea/DownloadAsset.aspx?id=14798).

[14] This accounting is for the research function only and excludes funding for the Cooperative Extension system.

[15] Keith O. Fuglie and Paul W. Heisey, Economic Returns to Public Agricultural Research, USDA Economic Research Service, Economic Brief Number 10, Washington, DC, September 2007, http://www.ers.usda.gov/Publications/EB10/.

[16] J. M. Alston, C. Chan-Kang, and M. C. Marra et al., A Meta-Analysis of Rates of Return to Agricultural R&D, International Food Policy Research Institute, 2000, http://www.ifpri.org/sites/default/files/publications/rr113.pdf.

[17] Philip Pardey, "Putting U.S. Agricultural R&D and Productivity Developments in Perspective," Farm Foundation Conference, 2009, http://www.farmfoundation.org/news/articlefiles/1705-Pardey%20.pdf.

[18] For example, see Pursuing a Unifying Message: Elevating Food, Agricultural and Natural Resources Research as a National Priority, Charles Valentine Riley Memorial Foundation, 2014 (http://192.254.250.185/~swcs/wp-content/uploads/2014/11/UnifyingReport_8Frevised 2.pdf); USDA Economic Research Service, Public Agriculture Research Spending and Future U.S. Agricultural Productivity Growth: Scenarios for 2010-2050, EB-

17, 2011 (www.ers.usda. gov/media/118663/eb17.pdf); and Food and Agriculture Organization of the United Nations, "2050: Increased Investment in Agricultural Research Essential," 2009 (http://www.fao.org/news/story/en/item/35686/icode).

[19] CRS Report R43938, FY2016 Agriculture and Related Agencies Appropriations: In Brief.

[20] USDA-REE, A Roadmap for USDA Science, 2010, at http://www.ree.usda.gov/ree/news/REE_Roadmap9_final.pdf.

[21] Create Research, Extension, and Teaching Excellence for the 21st Century (CREATE-21). APLU was previously known as National Association of State Universities and Land Grant Colleges (NASULGC); NASULGC changed its name to APLU in March 2009.

[22] USDA-REE, Research, Education, and Economics Action Plan, February 2012, revised March 2014, at http://www. ree.usda.gov/ree/news/USDA_REE_Action_Plan_03-2014.pdf.

[23] See footnote 18.

[24] President's Council of Advisors on Science and Technology, Report to the President on Agricultural Preparedness and the Agriculture Research Enterprise, Washington, DC, December 2012, http://www.whitehouse.gov/sites/default/ files/microsites/ostp/pcast_agriculture_20121207.pdf.

In: U.S. Nutrition and Agricultural Research ISBN: 978-1-63484-170-2
Editor: Arlene Ramos

Chapter 3

PUBLIC AGRICULTURAL RESEARCH SPENDING AND FUTURE U.S. AGRICULTURAL PRODUCTIVITY GROWTH: SCENARIOS FOR 2010-2050*

Paul Heisey, Sun Lin Wang and Keith Fuglie

MAJOR FINDINGS

- By 2050, global agricultural demand is projected to grow by 70-100 percent due to population growth, energy demands, and higher incomes in developing countries. Meeting this demand from existing agricultural resources will require raising global agricultural total factor productivity (TFP)[1] by a similar level. Maintaining the U.S. contribution to global food supply would also require a similar rise in U.S. agricultural TFP.
- TFP growth in U.S. agriculture is predicated on long-term investments in public agricultural research and development (R&D). Productivity growth also springs from agricultural extension, farmer education, rural infrastructure, private agricultural R&D, and technology

* This is an edited, reformatted and augmented version of Economic Brief Number 17, issued by the U.S. Department of Agriculture, Economic Research Service, July 2011.

transfers, but the force of these factors is compounded by public agricultural research.

- The rate of TFP growth (and therefore output growth) of U.S. agriculture has averaged about 1.5 percent annually over the past 50 years. Stagnant (inflation-adjusted) funding for public agricultural research since the 1980s may be causing agricultural TFP growth to slow down, although statistical analyses of productivity growth trends are inconclusive.
- ERS simulations indicate that if U.S. public agricultural R&D spending remains constant (in nominal terms) until 2050, the annual rate of agricultural TFP growth will fall to under 0.75 percent and U.S. agricultural output will increase by only 40 percent by 2050. Under this scenario, raising output beyond this level would require bringing more land, labor, capital, materials, and other resources into production.
- Additional public agricultural R&D spending would raise U.S. agricultural productivity and output growth. Raising R&D spending by 3.73 percent annually (offsetting the historical rate of inflation in research costs) would increase U.S. agricultural output by 73 percent by 2050. Raising R&D spending by 4.73 percent per year (1-percent annual growth in inflation-adjusted spending) would increase output by 83 percent by 2050.

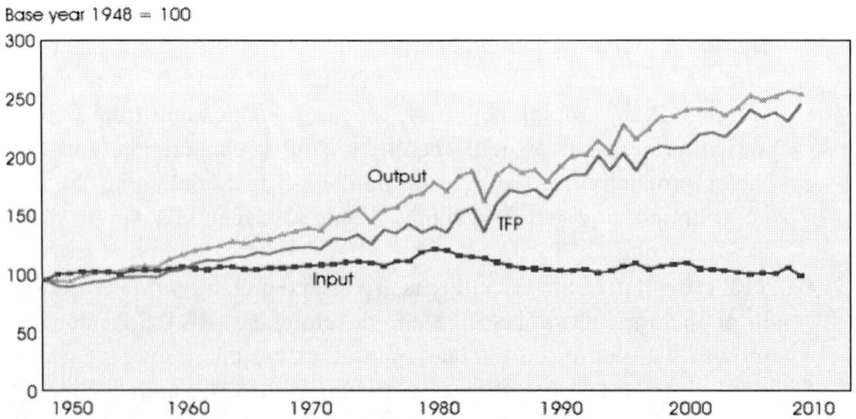

Source: U.S. Department of Agriculture, Economic Research Service.

Figure 1. U.S. agricultural output, input, and TFP indexes.

Productivity Growth Accounts for Nearly All U.S. Agricultural Growth

Total factor productivity (TFP) is a good indicator of technological change. It measures the efficiency with which all inputs (land, labor, capital, and materials) are combined to produce total outputs (all crop and livestock commodities). In U.S. agriculture, growth in TFP is nearly synonymous with growth in output since the overall size of the resource base has barely increased over the last 60 years. Between 1948 and 2008, the average annual growth rate in U.S. agricultural output was 1.58 percent and the average annual growth rate in TFP was 1.52 percent. As a result, total agricultural output in 2008 was 2.5 times that of 1948 (figure 1).

R&D Spending Spurs TFP Growth

Growth in TFP is strongly associated with the adoption of new technologies that raise yields or lower costs. Public investment in agricultural R&D is a major source of new agricultural technology; it also complements (raises returns to) other productivity-enhancing activities like extension, education, infrastructure, and private R&D. While private-sector investment in agricultural R&D also contributes new technology, much of this contribution is already accounted for in the way the inputs are measured in constructing the estimates of TFP.[2] Moreover, studies have shown that private R&D depends critically on government and university investments in science and technology.[3] Thus, even if the quality-adjusted measures of TFP do not fully net out the contributions of private R&D, if private R&D follows (responds positively to) trends in public R&D, then simulations using public R&D as the primary policy lever for future productivity growth should not be unduly biased. Nonetheless, the simulations reported here depend on these assumptions about private R&D: that its impacts are partially netted out of the TFP measure and that it follows trends in public R&D.

Statistical models of TFP growth treat R&D spending as "knowledge capital." As with physical capital, an R&D investment can affect productivity growth for several years (even decades), but eventually depreciates as technologies grow ineffective (e.g., new pests and diseases evolve) or obsolete. R&D investments typically begin boosting TFP within 3-5 years, with benefits peaking after 10 to 20 years, and with some impacts lasting as long as 50 years (Huffman and Evenson, 2006; Alston et al., 2010).

AGRICULTURAL R&D SPENDING SCENARIOS ACCOUNT FOR INFLATION IN RESEARCH COSTS

Between 1927 and 2009 (the period for which we have data), total spending by Federal and State agricultural research institutes on productivity-related agricultural research (excluding R&D on postharvest, environmental, and rural development issues) rose at an average annual rate of 7.49 percent in nominal terms. If adjusted by the Consumer Price Index, then research spending rose by 4.37 percent per year. However, if inflation adjustments reflect the cost of doing research (i.e., scientist salaries, laboratory equipment, etc.), then real research spending rose by only 3.03 percent per year (i.e., research costs have been rising faster than the average price level of goods and services in the economy).

Estimates of R&D knowledge capital are based on real research spending (adjusted by the rate of inflation in the cost of research resources). Growth in public agricultural research spending slowed dramatically after the early 1980s and has declined by more than 20 percent since peaking in 1994 (figure 2). The growth in the stock of agricultural knowledge capital, which follows trends in research spending with a lag, peaked around 2005. Since then, knowledge capital stock has started to decline as current spending (creation of new knowledge capital) has failed to keep up with knowledge capital depreciation.

We simulate three possible scenarios for future public agricultural R&D expenditure for 2010-2050 (see Box for a description of the simulation model). These scenarios include total expenditures on productivity-oriented agricultural research by USDA research agencies and State agricultural experiment and veterinary medicine colleges from all sources of funding (USDA, other Federal and State governments, and others).

Scenario 1: Public research spending is constant in nominal terms at the 2005-09 average ($2.5 billion per year). This implies a decline in real research expenditure by 3.73 percent per year (the inflation rate in the cost of research resources over 1983-2009).[4] In other words, under this scenario, inflation reduces the effective amount of research effort.

Scenario 2: Public research spending is constant in real terms at the 2005-09 average. This implies that nominal expenditures for public agricultural research rise by about 3.73 percent per year to offset inflation in the cost of research.

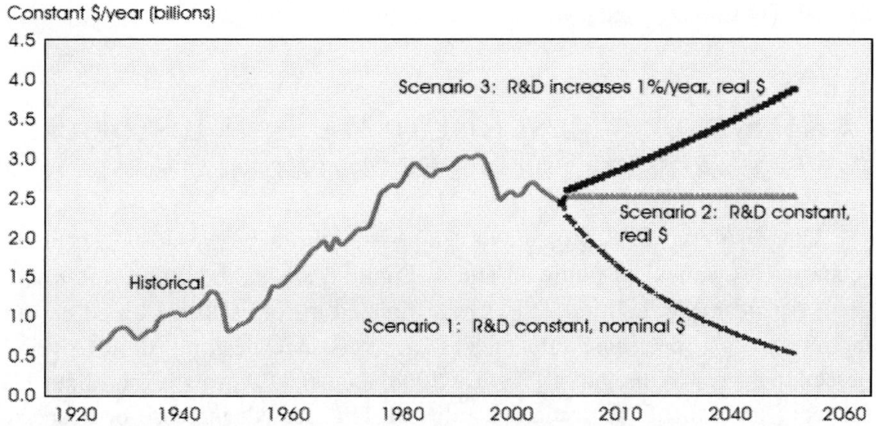

Source: Historical data from Huffman (2010) and 2010-2050 data from U.S. Department of Agriculture, Economic Research Service simulations.

Figure 2. Productivity-oriented agricultural research expenditures.

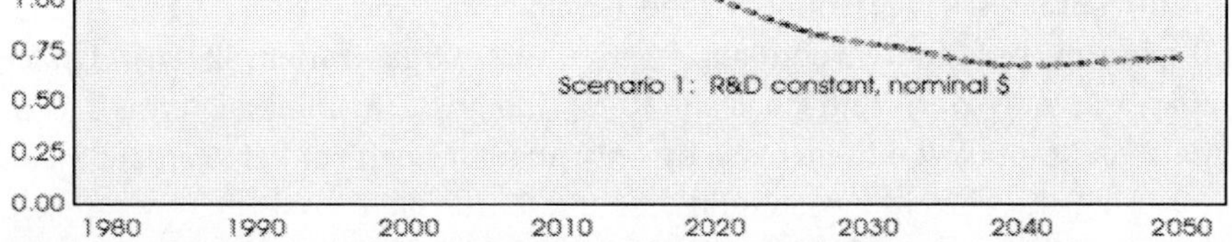

* The graph shows the TFP growth rate predicted from the statistical model rather than the actual TFP growth rate for 1980-2008. Actual TFP growth is highly variable, ranging between -15% and + 15% for individual years. This variability is mainly due to weather.

Figure 3. TFP growth rate projections.

Scenario 3: Public research spending increases by 1 percent per year in real terms (implying a nominal increase of about 4.73 percent per year). This 1-percent real growth rate is roughly equivalent to the rate of U.S. population growth (a proxy for growth in domestic agricultural demand). Real research expenditures rise, but more slowly than the rapid growth in real spending observed before the 1980s.

R&D SPENDING MUST RISE TO MAINTAIN HISTORICAL AGRICULTURAL TFP GROWTH RATE

The resulting TFP projections under the three public R&D spending scenarios are shown in figures 3 and 4. Figure 3 shows the projected *annual growth rate* in agricultural TFP. The estimated historical trend based on our model shows a declining rate of TFP growth since peaking at just over 2 percent per year in the mid-1980s. By 2008, annual TFP growth was about 1.3 percent.[5] In each of the scenarios, rates of TFP growth increase or decrease from present levels and then level off after 2030 or 2040. Under scenario 1, the TFP growth rate declines to under 0.75 percent per year. Under scenario 2, TFP growth stabilizes at about 1.4 percent per year. Under scenario 3, TFP growth accelerates and reaches 1.6 percent per year by 2040.

Figure 4 translates TFP growth into an agricultural TFP index (base year of 2008=100) under alternative funding scenarios. It also shows the agricultural output level (relative to 2008 production) achievable if agriculture's current resource base remains unchanged. Under scenario 3, by 2050 U.S. agricultural output would be 83 percent higher than 2008 levels and under scenario 2, output would grow by 73 percent. As a point of comparison, current estimates suggest global food demand will increase between 70 and 100 percent by 2050 (Food and Agriculture Organization, 2006). Under scenario 1, however, with constant R&D funding (in nominal terms), U.S. agricultural output would increase by only about 43 percent between 2008 and 2050 using the current agricultural resource base. This would barely outpace the projected growth in U.S. population. By contrast, U.S. agricultural output would likely keep up with the growth in global demand under the higher R&D investment scenarios.

Another recent study (Alston et al. 2010) used different data and modeling assumptions to independently derive statistical relationships between public agricultural R&D spending and TFP growth. Their findings are broadly similar to our own: public spending on agricultural R&D will need to rise to maintain historical rates of productivity growth in U.S. agriculture.

Implications of a decline in the long-term growth rate in U.S. agricultural productivity are significant for both the United States and the world. Slower productivity growth will likely cause agricultural prices to rise, which would reduce global economic welfare and raise poverty in urban areas and among food-deficit rural households, especially in developing countries (Nelson et al., 2010). Rising prices would also increase pressure to expand agricultural cropland at home and abroad and input use on cropland would likely intensify. Such agricultural resource expansion and intensification could come with significant environmental costs, such as further impairment of soil and water quality, loss of biodiversity, and increased emissions of greenhouse gases (Foresight, 2011). Finally, agricultural exports may also decline as U.S. farmers lose competitiveness in international markets (Ball et al., 2010).

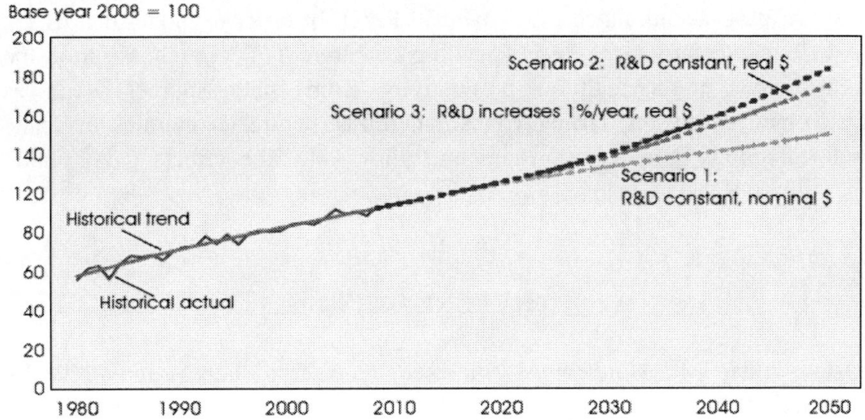

Figure 4. TFP index projections.

How the Estimates Were Derived

We use a model developed by ERS (Wang et al., 2010) along with research expenditures data from Wallace Huffman at Iowa State University. Huffman isolates the component of total public agricultural R&D associated with raising crop and livestock productivity and excludes other kinds of R&D (Huffman et al., 2001; Huffman and Evenson, 2006; Huffman, 2010).

Statistical models that link TFP to these investments treat R&D spending as "knowledge capital." Like physical capital, knowledge capital contributes to growth over a long period of time but eventually depreciates; it requires new R&D investment to be maintained and expanded. We use the assumptions developed by Huffman and colleagues to convert annual R&D investments, or flows, into R&D capital, or stocks. Under these assumptions, R&D investments influence R&D stocks, and therefore TFP, for up to 35 years.

We assume alternative scenarios for future public R&D funding to simulate future growth in U.S. agricultural total factor productivity to 2050. We use statistical relationships based on different productivity and investment patterns across States over 1980-2004 to estimate the effects of R&D stocks on productivity growth. These estimated relationships are used to project future TFP growth patterns given alternative assumptions about future R&D spending. Changes in current levels of public R&D spending affect future TFP growth only gradually since most of today's "knowledge capital stock" is the result of past, accumulated investment in R&D. In order to model the effects of different scenarios for R&D spending on future TFP growth, we hold the contributions to agricultural productivity from other sources—such as agricultural extension, farmer education, infrastructure, economies of scale, and technology transfer—constant and allow only the effects from public agricultural R&D to change.

REFERENCES

Alston, Julian M., Matthew A. Andersen, Jennifer S. James, and Philip G. Pardey. 2010. Persistence Pays: U.S. Agricultural Productivity Growth and the Benefits from Public R&D Spending. Springer.

Ball, V. Eldon, Jean-Pierre Butault, Carlos San Juan, Ricardo Mora. 2010. "Productivity and international competitiveness of agriculture in the European Union and the United States." Agricultural Economics 41: 611-627.

Economic Research Service. "Data Set: Agricultural Productivity in the United States," Agricultural Research and Productivity Briefing Room. Available at http://www.ers.usda.gov/Data/AgProductivity/ (accessed June 1, 2011).

Food and Agriculture Organization. 2006. World agriculture: towards 2030/2050, Interim report. Food and Agriculture Organization of the United Nations, Rome, June. http://www.fao.org/docrep/ oo9/a0607e/ a0607e00.htm

Foresight. 2011. The Future of Food and Farming: Challenges and Choices for Global Sustainability. Final Project Report. The Government Office for Science, London, United Kingdom.

Fuglie, Keith O. and Thomas S. Walker. 2001. "Economic Incentives and Resource Allocation in U.S. Public and Private Plant Breeding. Journal of Agricultural and Applied Economics 33 (3): 459-473.

Huffman, W. E., A. McCunn, and J. Xu. 2001. "Public Agricultural Research Expenditures with an Agricultural Productivity Emphasis: Data for 48 States, 1927-1995." Iowa State University, Department of Economics, Staff Paper.

Huffman, W. E., and R.E. Evenson. 2006. Science for Agriculture—A Long-Term Perspective. Second edition, Blackwell Publishing.

Huffman, W.E. 2010. "Measuring Public Agricultural Research Capital and Its Impact on State Agricultural Productivity in the U.S." Iowa State University, Department of Economics, Mimeo. (Updates data from Huffman et al., 2001)

Nelson, Gerald C., Mark W. Rosegrant, Amanda Palazzo, Ian Gray, Christina Ingersoll, Richard Robertson, Simla Tokgoz, Tingju Zhu, Timothy B. Sulser, Claudia Ringler, Siwa Msangi and Liangzhi You. 2010. Food Security, Farming and Climate Change to 2050. International Food Policy Research Institute, Washington, DC.

Wang, Chenggang, Yin Xia, and Steven Buccola. 2009. "Public Investment and Industry Incentives in Life-Science Research." American Journal of Agricultural Economics 91(2): 374-388.

Wang, Sun Ling, Eldon Ball, Lilyan Fulginiti, and Alejandro Plastina. 2010. "Benefits of Public Research Spill-in, Extension, and Road Networks in U.S. Agriculture." (Under revise and resubmit with a peer reviewed journal).

End Notes

[1] Total factor productivity (TFP) is the broadest measure of productivity. It compares the total output of a sector to the total land, labor, capital, and material inputs used to produce that output. Increases in TFP imply more output is forthcoming from a given level of inputs, or, equivalently, fewer inputs are required to produce the same output. Growth in TFP is considered to be an indicator of the rate of technical change in a sector.

[2] New technologies from private R&D are partially embodied in the quality of the inputs (chemicals, machinery, seed, etc.) purchased by farmers. The TFP measure used in this

study adjusts for such quality improvements in these inputs. In other words, it already nets out many of the productivity effects of private R&D.

[3] Public-sector research opens up new technological opportunities for the private sector to commercialize. One recent study found that "government expenditures in both basic biological research and agricultural and medical science create substantial spillovers for private firms...Indeed, opportunities created through public research are the principal source of growth in industry life sciences" (Wang, Xia and Buccola, 2009). Fuglie and Walker (2001) also found complementarity between public and private investments in crop genetic improvement, with public research in basic plant breeding spurring more private investment in crop variety development.

[4] While the average rate of inflation in research costs over 1927-2009 was 4.46 percent, this has slowed in more recent years. To project research cost inflation in the future, we chose the average inflation rate in research costs during 1983-2009, or 3.73 percent per year.

[5] This apparent decline in TFP growth is based on the model used to project future TFP growth. All other factors were kept fixed except for the changes in R&D "knowledge capital" stock; therefore, the decline in the TFP growth rate reflects the erosion of R&D stocks, as hastened by the stagnation in R&D funding that began in the 1980s (see figure 2). Thus, the slowing of TFP growth in figure 3 after the mid-1980s reflects this model assumption; actual agricultural TFP growth may not have slowed if other contributing factors offset the decline in R&D capital.

Index

#

20th century, 8

A

accounting, 19, 59
acid, 12, 15
ADA, 37
adaptation, 52, 58
administrators, 29, 30
adolescents, 8
adults, 6, 7, 8
age, 8, 41
agencies, 5, 11, 18, 19, 20, 22, 24, 32, 34, 36, 43, 44, 46, 48, 51, 54, 55, 64
agricultural economics, 46
agricultural exports, 67
agricultural market, 45
Agricultural Research Service (ARS), 43, 44, 47, 48
agricultural sector, 52, 55, 57, 58
agriculture, 43, 44, 45, 47, 51, 57, 58, 59, 60, 61, 62, 63, 66, 67, 68
alcohols, 26
amino, 26
amino acid(s), 26
anemia, 7
annual rate, 16, 20, 37, 62, 64
anorexia, 3, 7, 25, 26
appetite, 25, 26
appropriations, 5, 20, 24, 46, 47, 48, 49, 50, 51
ARS, 44, 47, 48, 52, 55
authority, 44, 46, 53

B

base, 63, 66
base year, 66
basic research, 5, 42, 47, 50, 51
behaviors, 5, 6, 23, 27
benefits, 2, 5, 8, 11, 12, 16, 18, 41, 58, 64
beverages, 16
bioavailability, 26
biodiversity, 67
biomarkers, 5
brain, 15
brain damage, 15
breeding, 70
budget surplus, 54

C

caliber, 51
calorie, 15
cancer, 3, 26, 42
candidates, 51
capacity building, 49
carbohydrates, 3, 26

cardiovascular disease, 7, 26
categorization, 34
causal relationship, 12
CDC, 7, 15, 38
Census, 45, 49
challenges, 2, 5, 51, 55, 57, 58
chemical(s), 5, 69
Chicago, 8, 39
chicken, 8
childhood, 8
children, 8, 15, 41
cholesterol, 3, 12, 16
classification, 16, 25, 42
climate, 52, 55, 57, 58
climate change, 55, 57
coding, 34
colleges, 4, 17, 22, 31, 32, 33, 47, 50, 64
commercial, 23, 35
commodity, 56
communication, 6, 7, 33
community(s), 7, 40, 45, 46, 51
comparative analysis, 19
competitiveness, 57, 67
complementarity, 70
composition, 26, 27, 32
computer, 34
Congress, 20, 31, 43, 44, 47, 49, 54, 58
conservation, 44, 52, 57
constituents, 3
construction, 8, 42
consulting, 6
Consumer Price Index, 64
consumers, 2, 5, 6, 8, 12, 15, 18
consumption, 5, 12, 15, 16, 27
consumption habits, 5, 15
consumption patterns, 27
Continuing Survey of Food Intakes by Individuals, 12
cooperative agreements, 22
coordination, 39, 47
coronary heart disease, 12, 16
cost, 11, 12, 19, 36, 64, 65, 70
cotton, 58
creativity, 23, 38
crop(s), 44, 45, 47, 56, 57, 58, 63, 67, 70
crop insurance, 58
curriculum, 39

D

database, vii, 1, 2, 4, 18, 19, 32, 34, 35, 42
deaths, 15
defects, 15
deficiency, 15
deficit, 59, 67
Department of Defense, 18, 34
Department of Health and Human Services, 1, 2, 8, 18, 41
depreciation, 64
destiny, 39
developing countries, 61, 67
diabetes, 3, 12, 26
Diabetes, 22
diet, 5, 6, 8, 12
Dietary Guidelines, 8, 10, 12, 41
Dietary Guidelines for Americans, 8, 12
dietary intake, 12, 15
dietary issues, 15
disclosure, 39
diseases, 3, 7, 26, 29, 42, 63
distribution, 4, 19, 24, 25, 31, 48, 51
District of Columbia, 31
drawing, 12

E

economic growth, 8, 56, 57
Economic Research Service (ERS), 1, 7, 17, 21, 24, 28, 29, 31, 32, 33, 36, 37, 38, 39, 40, 41, 43, 45, 46, 54, 55, 59, 61, 62, 65, 68
economics, 17, 18, 38
economies of scale, 68
education, vii, 1, 5, 17, 25, 27, 39, 43, 47, 50, 52, 57, 58, 61, 63, 68
educational materials, 11
educational programs, 45
egg, 47
election, 39

employees, 44, 45, 48
energy, 3, 12, 58, 61
environment(s), 40, 45, 46
Environmental Protection Agency, 8
environmental sustainability, 58
epidemiology, 5
equipment, 64
erosion, 70
ERS, 12, 45, 48, 55, 62, 67
European Union, 68
evidence, 5, 8, 10, 12, 38
evidence-based policy, 5
evolution, 18
exclusion, 47
expenditures, 19, 20, 21, 52, 64, 65, 66, 67
extreme weather events, 57

F

Farm Bill, 58
farmers, 58, 67, 69
fat, 6, 12, 15
fatty acids, 12
federal funds, 44, 52
Federal Government, 12, 15, 22, 28, 35, 42, 58
Federal Register, 41
fiber, 15, 26, 52
financial, 4, 17, 19, 20, 35, 36
financial data, 19, 35, 36
financial support, 17, 19
fiscal year, 4, 18, 36, 41
flexibility, 23
floods, 57
flour, 15
folic acid, 15
food, vii, 1, 2, 3, 5, 6, 11, 12, 15, 16, 17, 26, 27, 30, 32, 41, 42, 44, 45, 46, 52, 57, 58, 61, 66, 67
food additive(s), 12, 16
Food and Drug Administration (FDA), 11, 12, 15, 16, 41
food production, 12
food products, 2, 5, 15
food safety, 44, 46, 57
force, 51, 62
formula, 2, 3, 22, 23, 29, 31, 32, 34, 39, 42, 43, 44, 46, 49, 50, 51, 52
foundations, 17, 31, 32
freedom, 37
fruits, 16
functional food, 16, 17, 40
funding, 2, 3, 5, 17, 18, 19, 20, 22, 23, 24, 29, 30, 31, 32, 34, 35, 36, 42, 43, 45, 46, 47, 48, 49, 50, 51, 52, 54, 55, 57, 58, 59, 62, 64, 66, 68, 70
funds, 22, 31, 37, 39, 43, 44, 45, 46, 48, 49, 50, 51

G

GDP, 8
General Accounting Office (GAO), 17, 39
genes, 5, 39
genetics, 26, 29
genome, 44
global demand, 66
global economic welfare, 67
goods and services, 8, 64
government expenditure, 70
government policy, 27
governments, 18, 22, 31, 41, 55, 64
granola, 16
grants, 1, 3, 22, 23, 43, 44, 45, 46, 48, 49, 50, 51, 52
graph, 65
greenhouse, 67
greenhouse gas(s), 67
gross domestic product, 8
growth, 16, 20, 29, 35, 40, 56, 57, 61, 62, 63, 64, 65, 66, 67, 68, 70
growth rate, 35, 63, 65, 66, 67, 70
guidance, 11

H

HAART, 27
hazardous waste, 8

health, 2, 5, 7, 8, 10, 12, 15, 16, 17, 33, 40, 42, 46, 57, 58
health care, 7
health care costs, 7
health effects, 8, 12
health problems, 15
health risks, 12
health status, 8
heart disease, 12
HHS, 2, 8, 18
higher education, 17, 44, 50
highly active antiretroviral therapy, 27
HIV, 27
HIV/AIDS, 27
House, 47
housing, 8
human, 3, 6, 8, 12, 16, 17, 20, 24, 25, 27, 39, 41, 42, 44, 45, 52, 57, 58
human body, 3
human health, 6, 8, 12, 17, 45, 57
hydrogenation, 12

I

identification, 18, 35
improvements, 8, 16, 39, 57
income, 57, 58
income support, 58
individuals, 5, 6, 8, 10, 12, 50
industry(s), 8, 16, 17, 40, 52, 54, 70
infection, 8, 26
inferences, 17
inflation, 42, 54, 57, 62, 64, 65, 70
infrastructure, 61, 63, 68
ingredients, 27
initiation, 31
injury(s), 8, 26
institutions, 2, 3, 4, 16, 17, 30, 31, 32, 33, 34, 42, 45, 47, 48, 50, 51, 52, 59
insulin, 12
insulin resistance, 12
international competitiveness, 68
intervention, 12, 15
investment(s), 1, 2, 4, 17, 18, 20, 34, 56, 61, 63, 66, 68, 70
iodine, 15
Iowa, 67, 69
iron, 15
irrigation, 57
issues, vii, 33, 43, 47, 50, 64

K

kidney, 42
knowledge capital, 63, 64, 68, 70

L

labeling, 12
LDL, 12, 16, 38
lead, 20, 42
learning, 2, 5
legislation, 42
leisure, 8
life cycle, 26
life expectancy, 8
life sciences, 38, 70
lifetime, 8
lipids, 26
liver, 42
livestock, 44, 47, 63, 67
longevity, 8

M

machinery, 69
magnitude, 12
majority, 47, 48, 55
malnutrition, 2, 5
management, 57
manufacturing, 8, 12
market segment, 17
marketing, 57
Maryland, 22
mass, 6
materials, 11, 62, 63
meat, 47
media, 54, 56, 60
medical, 4, 10, 15, 16, 26, 31, 32, 70

medical science, 70
medicine, 47, 48, 64
melting, 12
membranes, 16
mental health, 42
messages, 11
Metabolic, 3, 26, 28, 30
metabolic responses, 5
metabolism, 15, 26
metabolites, 5
micrograms, 15
micronutrients, 12, 15
mission(s), 5, 22, 24, 31, 42, 43, 44, 47, 51
models, 63, 68
mortality, 8

N

National Academy of Sciences (NAS), 12, 40, 50, 51, 59
National Aeronautics and Space Administration, 18
National Institutes of Health, vii, 1, 2, 4, 18, 20, 35, 36, 40, 51
National Research Council, 40
National School Lunch Program, 11
natural resource management, 44, 45, 55, 57
natural resources, 44, 46
nonprofit organizations, 17
NRC, 40
nutrient(s), 5, 11, 12, 15, 26, 27
nutrition, vii, 1, 2, 3, 4, 5, 6, 7, 8, 10, 11, 12, 15, 16, 17, 18, 19, 20, 21, 22, 23, 24, 25, 26, 27, 28, 29, 30, 31, 32, 33, 34, 35, 36, 37, 39, 40, 41, 42, 44, 46, 52, 57
nutritional deficiencies, 15
nutritional status, 27

O

obesity, 2, 3, 5, 6, 7, 18, 25, 26, 29, 38, 39, 41
obesity prevention, 25
opportunities, 22, 29, 34, 40, 47, 57, 70
osteoporosis, 26
oversight, 47

P

patents, 4
pathogens, 58
peer review, 69
performers, 5, 24
personal communication, 55
pests, 58, 63
physical activity, 11
phytosterols, 16
policy, 4, 17, 27, 34, 45, 47, 52, 63
policymakers, 8, 57
polyunsaturated fat, 12
polyunsaturated fatty acids, 12
population, 6, 8, 12, 57, 61, 66
population growth, 57, 61, 66
portfolio, 1, 2, 3, 25, 27, 28, 29, 30, 31, 32, 58
poultry, 47
poverty, 67
preschool, 25
preschool children, 25
preservation, 3
President, 40, 51, 59, 60
prevention, 26, 39
private firms, 70
Private Funding, 52
private investment, 70
private portfolio, 39
private sector, 17, 44, 55, 58, 70
procurement, 22
product attributes, 8
product market, 8
production costs, 57
productivity growth, vii, 62, 63, 67, 68
profit, 32
profitability, 57
project, 4, 18, 19, 20, 23, 24, 25, 26, 27, 28, 34, 35, 36, 37, 41, 68, 70
protection, 44
proteins, 3, 5, 26, 31, 42
public goods, 18

public health, 2, 5, 7
public investment, 2, 57
public sector, 8
public-private partnerships, 54
Puerto Rico, 45, 47

Q

quality improvement, 70
quality of life, 8

R

R&D investments, 17, 34, 63, 68
rate of return, 56
real terms, 1, 2, 65, 66
reality, 25
recommendations, 8, 10, 11, 51, 57
regional problem, 44
regression, 37
regression analysis, 37
regulations, 5, 6, 11, 15
relevance, 18, 25, 26, 27
reliability, 51
renewable energy, 46
repair, 42
requirement(s), 12, 15, 41, 46, 49, 59
research funding, 22, 24, 30, 49, 55, 58
research institutions, 32
researchers, 2, 4, 6, 22, 23, 29, 30, 32, 33, 34
resource management, 52
resources, 2, 5, 34, 50, 52, 58, 59, 61, 62, 64
response, 29
restrictions, 23
retail, 8, 11, 12
revenue, 16
risk(s), 6, 8, 12, 16, 44, 52, 57
risk management, 57
room temperature, 12
rules, 23, 42
rural development, 44, 45, 57, 64
rural population, 46, 49

S

safety, 5, 16
Salmonella, 8
saturated fat(s), 12
saturated fatty acids, 12
school, 4, 11, 31, 32, 47, 48
science, 10, 17, 26, 30, 38, 63
scientific knowledge, 39, 43, 44
scientific progress, 16
seed, 69
Senate, 47
serum, 16
services, 6, 8, 18, 35
shape, 23
shelf life, 12
showing, 12
simulation(s), 62, 63, 64, 65
SNAP, 11
social sciences, 44
sodium, 15
soybeans, 58
specialty crop, 47
spending, vii, 7, 52, 54, 55, 57, 58, 62, 63, 64, 65, 66, 68
spillovers, 70
stability, 28
stakeholder groups, 57
standardization, 34
state(s), 39, 43, 45, 46, 47, 49, 50, 51, 52, 54, 55, 58, 59
statistics, 19, 44, 45
statutes, 22, 47
stock, 64, 68, 70
structure, vii, 1, 2, 5, 21, 24, 33, 57
Sun, v, 61, 69
Supplemental Nutrition Assistance Program, 11
surveillance, 52

T

Task Force, 57, 59
technical change, 69

technological change, 63
technology(s), 3, 6, 16, 18, 26, 32, 46, 54, 57, 58, 61, 63, 68, 69
technology transfer, 54, 62, 68
territory, 31, 58
terrorism, 54
threats, 47, 58
thyroid, 15
Title V, 47
total factor productivity, 61, 68
trace elements, 26
trade, 57
training, 18, 19, 35, 42
transactions, 8
translation, 6
trauma, 26
treatment, 26, 42
tribal colleges, 47
triggers, 6, 41

U

U.S. Department of Commerce, 18
United Kingdom, 69
United Nations, 60, 68
United States, 2, 8, 15, 16, 34, 38, 46, 56, 59, 67, 68
universities, 4, 17, 22, 31, 32, 33, 47, 50, 51, 55, 59
urban, 58, 67
urban areas, 67

V

valuation, 8
variables, 26
variations, 6
varieties, 57
vegetables, 16
vitamins, 15, 26, 27, 31, 42

W

wages, 8
Washington, 39, 40, 45, 46, 59, 60, 69
water, 26, 52, 58, 67
water quality, 67
weight gain, 6, 41
well-being, 42, 45
WIC, 11, 41
Women, Infants, and Children, 11
workers, 8
World War I, 56

Y

yield, 35